MATERIALS SCIENCE RESEARCH

Volume 2

MATERIALS SCIENCE RESEARCH

RESEARCH

Volume 2

The Proceedings of the 1964 Southern Metals/Materials Conference on
Advances in Aerospace Materials, held April 16-17, 1964,
at Orlando, Florida, hosted by the Orlando Chapter
of the American Society of Metals

Edited by

Henry M. Otte

and

Saul R. Locke

℗

PLENUM PRESS
NEW YORK
1965

ISBN 978-1-4684-7452-7   ISBN 978-1-4684-7450-3 (eBook)
DOI 10.1007/978-1-4684-7450-3

Library of Congress Catalog Card Number 63-17645

©1965 Plenum Press
Softcover reprint of the hardcover 1st edition *1965*

A Division of Consultants Bureau Enterprises, Inc.
227 West 17th Street • New York, N. Y. 10011

FOREWORD

The challenges of space exploration are a great stimulus to our technologies today. Development of successful aerospace programs has required the best efforts of the scientist and engineer in almost every discipline. Not so long ago, it truly could be said that designers are trying to develop tomorrow's vehicles with yesterday's materials. Unfortunately, we find that the situation remains nearly the same today. The purpose of this conference was to identify materials, processes, and methods that show the greatest potential in future space technology and to define the gap between mission requirements and materials application.

Of the many properties of materials, the one in which the largest gap between fundamental understanding and practical application appears to exist is the mechanical property, particularly of crystalline materials. The emphasis on crystalline materials is a natural one. It is these materials which are used primarily when demands are placed on mechanical strength, especially at elevated temperatures. The advent of space exploration requires the utilization of materials in environments and under conditions that are a challenge to the resourcefulness and ingenuity of the scientist and engineer. The scientist can, as a result of the past thirty years' work, relate mechanical properties to the formation, motion, and interaction of individual crystalline defects, such as vacancies, interstitials, and dislocations. Furthermore, he can, by controlled preparation of his materials, confine his studies to those cases in which the concentration of crystal defects is conveniently low. The engineer, for practical reasons, uses materials that contain a very large number of crystal defects, in which case, it becomes exceedingly difficult, if indeed still meaningful, to speak of the properties of the material in terms of the behavior of individual defects.

In order to derive maximum technological benefit from scientific advances, a continual, complementary relationship between the efforts of the engineer and scientist must be established and encouraged. It is in this spirit that the present conference was organized.

The papers are grouped into two parts—those that deal with basic observations (Part I) and those concerned with applied problems (Part II). The first paper in Part I reviews the importance of grain boundaries in controlling various properties (including ductility) of a wide variety of ceramic materials. In refractory metal oxides, the deviation from stoichiometry, the nature and ionization state of the defects that predominate, and the mechanism of charge transport are also important factors that affect the properties of the material; their measurement is detailed in the second paper. Defects, such as dislocation, can be observed in metals by electron transmission microscopy, and two papers describe this technique used to study changes in dislocation configurations produced by radiation damage in molybdenum and by deformation in beryllium. Although electron transmission microscopy permits direct observation of dislocations, the volume of material examined at any one time is of the order of 10^{-10} cm^3; in contrast, X-ray methods generally examine volumes larger by a factor of 10^8, but the usual diffraction techniques require interpretation. The last two papers in Part I deal with the calculation of small-particle size, stacking faults, and local (or microscopic) and homogeneous (or macroscopic) strains from changes in the broadening, shape, and position of X-ray diffraction powder-pattern lines.

Part II begins with three papers on refractory metals. The first deals with the effect of processing variables (e.g., amount of cold work and carbon content) on the strength and recrystallization properties of molybdenum-TZM alloy sheet. This is followed by a paper comparing the ductility of tungsten and its alloys prepared by the arc cast process and by powder metallurgy methods. The third paper is on the ductility of chromium-rich, two-phase alloys prepared by liquid-phase sintering. Ductility is taken for granted in the report on a process for forming sandwich structures on complex contours without detrimental effects on the bond juncture between the core and face sheets. For lightweight materials where mechanical strength is unimportant, but high-temperature insulation (temperatures in excess of 4000°F in vacuum or other environments) is required, ceramic and metal foams can be successfully prepared, as described in the next paper. There then follows a paper on the relationship of manufacturing techniques to such properties as thermal shock, erosion resistance, and strength of solid ceramics (carbides, borides, and oxides); ceramic and oxide coatings are also considered. The preparation and properties of boron nitride is the subject of a separate article which also deals with possible applications of the material.

The diversity of materials and problems encountered in the aerospace industry is effectively illustrated by the juxtaposition of the papers on graphite and on glass microtape. The former paper reports on the improvement obtained in the oxidation resistance of graphite at high temperatures by alloying with selected additives. The latter describes how the hydrogen permeability of the resin phase of glass-fiber-reinforced plastics has been overcome at cryogenic temperatures by use of glass microtape. The final paper shows how data on materials can be evaluated for design concepts of aerospace applications, such as uncooled, nuclear rocket nozzles, and emphasizes some of the material problems.

The papers assembled in this book were presented at the Southern Metals/Materials Conference held in Orlando, Florida, on April 16 and 17, 1964. The Orlando Chapter of the American Society of Metals acted as host. It is a pleasure to acknowledge the efforts of the many individuals who contributed to the success of the conference by serving on the review board and in other capacities.

The Editors

COMMITTEE PERSONNEL

General Chairman

J. Richter, Martin-Marietta Corporation

Technical Program

S. R. Locke (Chairman), H. M. Otte (Co-chairman), J. M. Neff (Co-chairman), Martin-Marietta Corporation; R. E. Reed-Hill, University of Florida; T. G. Olsen, University of Miami; R. Hochman, Georgia Institute of Technology

Arrangements

M. Dyer, W. Ben Wimberly Company

Publicity

S. Maszy, Martin-Marietta Corporation

Keynote Speaker

Merrill A. Scheil, President, American Society of Metals

Dinner Speaker

Dr. Earl T. Hayes, Assistant Director, Office of the Director of Defense, Research, and Engineering

TECHNICAL REVIEW BOARD

CONTENTS

Part I

Fundamental Problems

Part II

Applied Research

Part I
Fundamental Problems

Grain Boundaries In Ceramic Materials

Thomas D. McGee

Iowa State University, Ames, Iowa

This paper reviews the importance of grain boundaries in controlling the properties of a wide variety of ceramic materials. It includes results of original research in the effect of grain boundaries on the ductility of cubic ionic solids as a function of crystallographic orientation through the use of optical birefringence and transmission electron microscopic techniques.

Theoretical models for grain boundaries are compared with actual boundaries. The theory of dislocation interaction with grain boundaries is discussed. The grain boundary model is a useful tool for comparing the work hardening of cubic ionic crystals with theoretical calculations.

The sintering of ceramic oxides is diffusion-controlled. The relative value of the diffusion constants for the surface, the volume, and the grain boundaries determine whether low porosity can be reached. Low porosity and fine grain size are essential to obtaining high strength and creep resistance. The impurity concentration at grain boundaries and in the bulk of the material can be used to control the sintering process. In the selection of impurities to control grain size and porosity, grain boundary diffusion mechanisms must be considered.

The hope of ductile ceramics hinges upon an understanding of grain boundaries. The presence of grain boundaries can cause microcracks and brittle fracture by acting as a dislocation barrier. The effectiveness of the boundary in inhibiting dislocation motion is a function of the orientation of the crystal systems. Cubic ionic crystals of sodium chloride structure do exhibit some ductility in polycrystalline form under controlled conditions. Thermal diffusivity of ceramic oxides at incandescent temperatures is a function of grain boundary concentration. Optical transparency of high density oxides is a function of grain boundary concentration. Magnetic properties of barium ferrite permanent ceramic magnets are dependent on grain boundaries to inhibit domain rotation. By contrast, soft ferrites have higher permeability when grain boundaries and porosity are at a minimum.

I. INTRODUCTION

This is a review paper which is intended to illustrate the importance of grain boundaries in controlling the properties of modern ceramic materials. By showing this importance and by surveying the theory of grain boundaries, the author hopes to stimulate interest and promote needed research, which will improve theoretical understanding and afford application to practical problems.

Metallurgists are fully aware of the extreme importance of grain boundaries and grain size of metals in controlling such

3

properties as yield point, hardenability, brittle–ductile transition temperature, impact resistance, creep resistance, etc. It will come as no surprise to them that grain size is also important in ceramic materials. The extent of this importance may be surprising to some, however.

Many ceramic systems of technological importance have more than one phase present. Often one of the phases is a gas phase, i.e., a pore phase. This is especially true when lightweight, pure oxides are used, and where refractoriness and weight considerations are dominant.

Grain boundaries can exist as homogeneous boundaries, i.e., boundaries between crystals of the same phase, and as heterogeneous boundaries, i.e., boundaries between phases. Both types of boundaries are interfaces and the physical chemistry of interfaces is applicable. A surface is a special case of a heterogeneous grain boundary. A liquid or a gas is the second phase.

The term "grain" is used here in its usual metallurgical and ceramic connotation. A grain is a region in which a single general continuous crystal orientation exists. Grain boundaries may exist as subgrain boundaries or low-angle grain boundaries. These are special cases of the grain boundary which are given separate designation because of their importance. The usual grain is understood to be an imperfect crystalline region which will include more or less subgrain boundaries, i.e., regions which differ in orientation by extremely small angles—so small as to be difficult to measure.

Low-angle grain boundaries are those in which the dislocation theory of grain boundaries is applicable. As will be discussed later, the upper angular limit varies with the ceramic.

II. THEORY OF GRAIN BOUNDARIES

Five degrees of freedom exist for any grain boundary. One may specify two for the plane of the boundary and three for the crystal axes. Actual grain boundaries are not planar, however (Fig. 1). This is one of the many deviations from theory which may exist.

Low-Angle Grain Boundaries

The dislocation theory of grain boundaries has been well developed during the last fifteen years. An excellent summary of the knowledge in this area is given by Amelinckx and Dekeyser [1]. In preparing this discussion the author has relied heavily on their

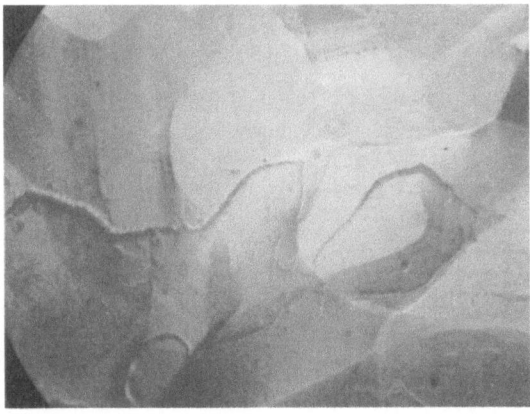

Fig. 1. Grain boundary of sodium chloride (×100).

summary.   Earlier works of importance which have contributed to this theory include those of Burgers [2], Bragg [3], Shockley and Read [4], and Van der Merwe [5]. A good introduction to the dislocation theory of grain boundaries is found in a book by Read [6].

Low-angle boundaries can be constructed from arrays of tilt and twist dislocations (Figs. 2 and 3).   Ceramic materials are usually compounds.   To maintain electrical neutrality, the structure must accommodate two extra planes of atoms, instead of one, as usually occurs in the case of an edge dislocation in a metal.

Some ionic ceramic materials have sodium chloride, calcium fluoride, or cesium bromide structures (Figs. 4, 5, 6, and 7). Many ceramic crystals have close-packed structures.   (The sodium chloride structure consists of two interpenetrating face-centered cubic lattices.)   One of the most common is the close-packed oxygen structure, such as found in alumina (hcp) and ferrite spinel (fcc). The distribution of cations in a close-packed anion structure is often complex.   The dislocation structure and the low-angle grain boundary configuration may be unique for these complex structures [7,8].

Multilayer crystalline structures also exist which may have unique configurations.   Some of these ceramic structures are characterized by mica, kaolinite, graphite, and molybdenum disilicide.   The analysis of dislocations in these in terms of stacking faults is parallel to that of many metals [9] (Fig. 8).

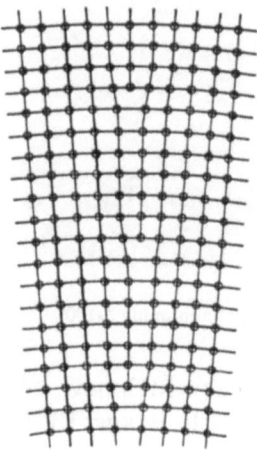

Fig. 2. Simple tilt boundary (Amelinckx and Dekeyser [1]).

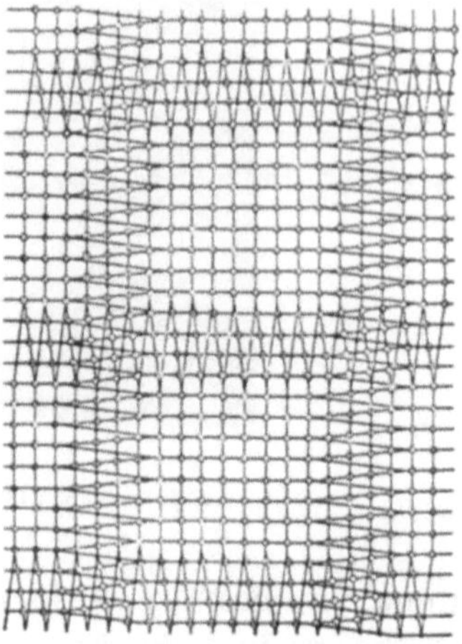

Fig. 3. A simple twist boundary
(Read [6]).

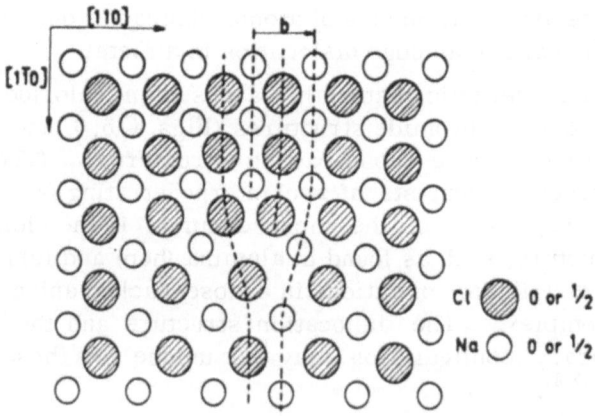

Fig. 4. Edge dislocation in sodium chloride (Amelinckx [57]).

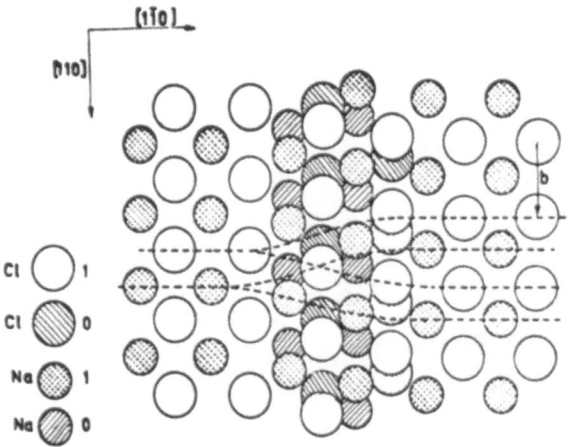

Fig. 5. Screw dislocation in sodium chloride (Amelinckx [57]).

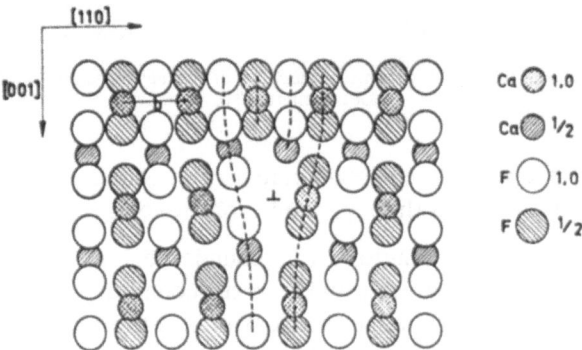

Fig. 6. Edge dislocation in calcium fluoride (Amelinckx [57]).

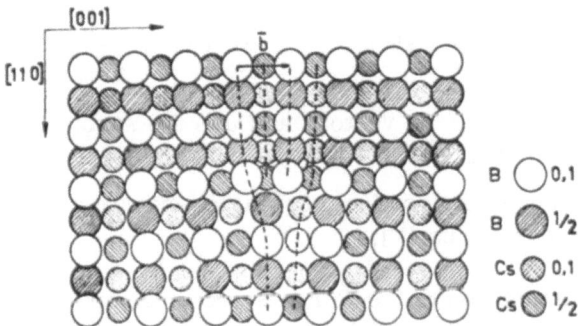

Fig. 7. Edge dislocation in cesium bromide (Amelinckx [57]).

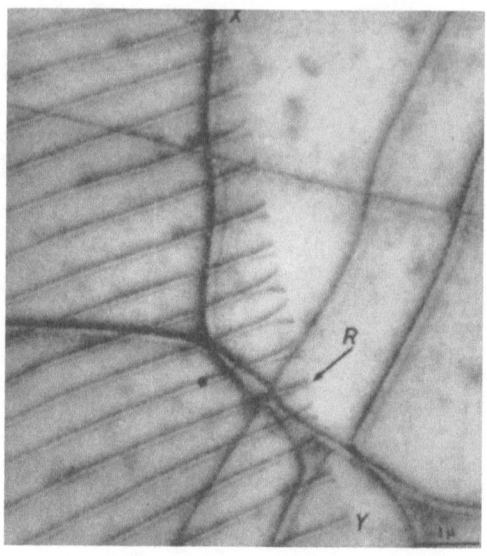

Fig. 8. Dislocation ribbons in graphite emerging at a cleavage step XY (Amelinckx and Delavignette [9]).

Certain properties of low-angle grain boundaries can be deduced from dislocation theory. Some of these are:

1. In general low-angle grain boundaries may be considered to be combinations of simple twist or tilt boundaries. The spacings of an array of parallel edge dislocations in the tilt component and spacing of the grid of screw dislocations in the twist component can be calculated by the following

$$\text{For tilt, } \frac{b}{d} = \sin\theta \qquad \text{For twist, } \frac{b}{2d} = \sin\theta$$

where b is the Burgers vector, and d is the spacing, and θ is the angle of twist or tilt [10,11].

2. Low-angle dislocation boundaries may be analyzed through the following conditions [12]:

a. The summation of the Burgers vector components normal to the boundary, multiplied by their respective dislocation density, determines any tilt component which may exist.

b. The summation of the Burgers vector components of the edge dislocations in the plane of the boundary, multiplied by their respective dislocation density, must be zero.

c. The summation of the Burgers vector components of the screw dislocations in the plane of the boundary, multiplied by their respective densities, is equal to twice the negative angle of twist.

3. Because of the elastic strain about a dislocation, certain stable configurations will exist which comply with 1 and 2 above. Other dislocations in the crystal will interact with these grain boundaries in a special way, because the dislocations of the boundary are grouped in a particular stable configuration. Thus the usual forces between individual dislocations are modified and may be increased or decreased depending on the boundary configuration. The following conclusions result:

a. Very small-angle grain boundaries may be mobile and move under stress.

b. In general, other dislocations moving toward a boundary are repelled by it if they lie in the same glide plane. The probability of passing a boundary decreases with increasing angle of tilt or twist, because the dislocations of the boundary become closer together. For an infinite simple tilt boundary, the force acting on an edge dislocation on a plane midway between two dislocations of the boundary is given by:

$$F_x = \frac{-Gb^2\pi^2 x}{2\pi(1-\nu)h^2\left[\coth(\pi x/h)\right]^2}$$

where G is the modulus of rigidity, ν is Poisson's ratio, x is the distance from the boundary, and h is the spacing between dislocations of the boundary.

c. Refinements in these calculations to include factors such as mobile and pinned boundary dislocations, two or more approaching dislocations, and mixtures of free and pinned dislocation have not changed the basic concepts [13].

d. The tensile and compressive strain fields of dislocations in grain boundaries attract impurity ions of larger and smaller size, respectively.

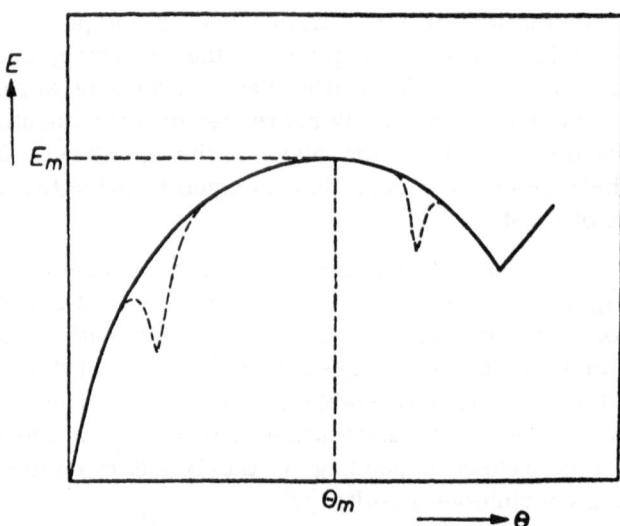

Fig. 9. Grain boundary energy as a function of angle; energy cusps shown as dotted lines (Van Buren [11]).

e. The energy of a grain boundary is a function of the angle of misfit. At certain angles atomic misfit is less than at other nearby angles. An energy cusp may result (Fig. 9). The energy grain boundary is not easy to calculate although special configurations have been calculated [14]. The expression for a simple tilt boundary is typical:

$$E = \frac{Gb\,\theta}{4\pi\,(1-\nu)}\;(A - \ln\theta)$$

Here θ is the angle of tilt, and A is a constant which equals $1 + \ln\,(b/2r_0)$, where r_0 is the radius of the dislocation core [1]. The core is defined as the central region in which the theory of elasticity is not applicable, and thus the angle at which the cores "touch" is unknown. The upper angular limit in which dislocation theory applies, the maximum low-angle grain boundary, depends upon the crystalline structure, the "softness" of the atoms or ions involved and the impurities present. Usually the angle is considered in the order of 15° to 25°. However the energies calculated from dislocation theory actually fit the experimental results over a much wider region [15].

High-Angle Grain Boundaries

Models of high-angle grain boundaries, proposed by Beilby [16], Mott [17], Shockley [18], Chalmers [19], and Li [20], are the amorphous layer, the islands of misfit, the dense array, the successive small-angle boundaries, and the dislocation core models, respectively. Li has given an analysis of the properties which a high-angle boundary should exhibit:

1. The model must reduce to the dislocation model at the low angles.
2. It must have viscous liquid behavior at high temperatures.
3. The model must produce a fairly constant energy over a range of high angles.
4. The activation energy for grain boundary shear must vary with angle, even at the high angles.
5. Grain boundary melting must occur above a given angle.
6. Diffusion, shear, and related phenomena must vary continuously with the angle.
7. The migration velocity must increase at higher angles and must be sensitive to impurities.

By an extension of Shockley's arguments, Li developed the dislocation core model, which is consistent with most of the above for a high-angle tilt boundary. He concludes that as the angle is increased and the dislocations come closer together, the model changes from a simple tilt model (Fig. 2) possibly to the Chalmers structure (Fig. 10) to an amorphous or liquid structure. Li believes that these changes are controlled by the dislocation cores which overlap at the high angles and produce viscous liquid structure (Fig. 11).

The principal arguments in favor of a viscous liquid structure at the high angles are the internal friction measurements on polycrystalline aluminum by Ke [21], the lower melting point of high-angle boundaries, and the unavailability of a suitable alternative structural model. But the decrease in the melting point is only a fraction of a degree in highly pure materials, and the term liquid implies flow which should occur at much lower temperatures. Obviously at very high angles, the atomic misfit must be very great. Unfortunately, no general theory for high-angle boundaries exists which satisfactorily describes the observed properties.

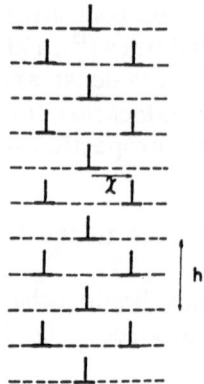

Fig. 10. Chalmers model of a medium-angle
tilt boundary (Amelinckx and Dekeyser [1]).

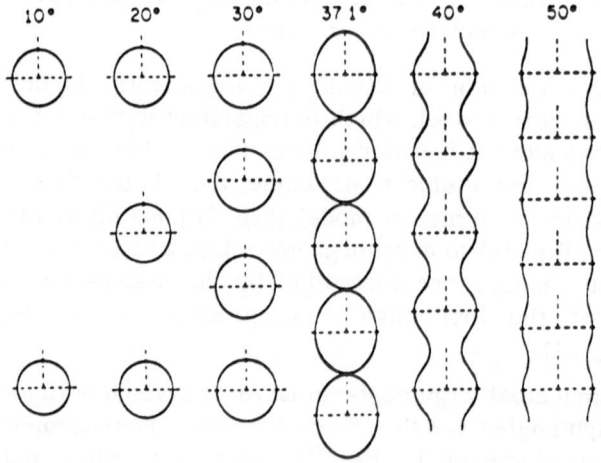

Fig. 11. The structure of a tilt boundary $r_0 = \dfrac{b}{2}$ (Li [20]).

This is particularly frustrating because actual grain boundaries in polycrystalline ceramics are usually high-angle grain boundaries.

Heterogeneous Grain Boundaries

The theory of heterogeneous grain boundaries is even less well defined than that of homogeneous grain boundaries. The principal theoretical efforts in this area consider the grain boundaries between lattices with slightly different lattice translation vectors in terms of the stresses and the grain boundary energy [5,22,23]. These employ either the Pierls–Nabarro expression or the parabolic expression for the forces in analyzing small-angle dislocation models corresponding to screw or edge configuration.

The results are in general agreement with the homogeneous small-angle dislocation theory. The forces normal to the interface were found to be small. The Pierls–Nabarro model was found to be the best representation for forces tangential to the interface. The strain energy was found to be highly localized, in a region within a radius of one-half the spacing between dislocations. No provision for polarization of the atoms was provided. The interfacial energy was found to be very sensitive to the dislocation spacing, rising to a high and fairly constant value at a low, but undefined, spacing. Although these conclusions are in general agreement with experiment, the theory is not well enough developed for application to real situations.

III. INFLUENCE OF GRAIN BOUNDARIES ON THE PROPERTIES OF CERAMIC MATERIALS

The dislocation model of a grain boundary leads to certain properties which can be subjected to theoretical analysis. An excellent discussion of properties is given in the article by Amelinckx and Dekeyser [1]. In this review the properties will be enumerated. Only those additional features of the theory which are more recent will be given full explanation.

Diffusion

Diffusional processes are of immense practical importance in the manufacture and use of ceramic materials. It is generally accepted that grain boundary diffusion is much more rapid than intragranular or bulk diffusion [24]. The ratio of the grain boundary diffusion "constant," D', to the bulk diffusion "constant," D, may be as high as 10^9 although the ratio is reduced as the temperature is increased. It appears possible, however, that diffusion might be decreased at a boundary if the strain might encourage compound formation.

The mathematical expression for grain boundary diffusion is quite complex. An excellent paper by LeClaire discusses the usual expressions which are often used for common diffusional experiments [25]. These experiments usually assume a high and constant source of the diffusing ion on one plane at any time, t. The progress of diffusion through material with a grain boundary of finite width, δ, for a distance y is a function which can be accurately described by an equation credited to Whipple [26]:

$$\frac{C}{C_0} = \operatorname{erfc} \tfrac{1}{2}\eta + \frac{\eta}{2\pi^{1/2}} \int_1^\infty \frac{d\sigma}{\sigma^{3/2}}\, e^{-\eta^2/4\sigma}\, \operatorname{erfc} \tfrac{1}{2}[(\sigma-1)/\beta + \xi]$$

where

$$\eta = \frac{y}{(Dt)^{1/2}}$$

$$\xi = \frac{x-\delta/2}{(Dt)^{1/2}}$$

$$\beta = \frac{D'\,(\delta/2)}{D\,(Dt)^{1/2}}$$

C is concentration and σ is a dummy variable.

In the above expression the first term gives the bulk diffusion contribution and the second gives the grain boundary contribution. Because of the complexity of the equation a simpler equation credited to Fisher [27] is usually used:

$$\frac{C}{C_0} = \exp\left(-\pi^{1/4}\eta\beta^{1/2}\right) \operatorname{erfc}(\xi/2)$$

Fisher's equation applies only if the contribution from direct bulk diffusion is negligible. This is true when β is large. It is especially dangerous to use Fisher's equation where low angles of tilt, long diffusion times, and high temperatures are involved. Improper use of Fisher's equation will lead to over estimating the activation energy for grain boundary diffusion and may result in significant error in $D'\delta$.

The usual results of diffusional experiments are that grain boundary diffusion in metals is very rapid and that δ is very small, often taken as about one lattice spacing at low angles, and two or three spacings at high angles. Usually, too, the activation energy for grain boundary diffusion is small and insensitive to angle of tilt or twist, although the higher-angle boundaries deliver more of the diffusing ion because of the closer spacing of dislocations.

In contrast to the above, Wuensch and Vasilos recently studied the diffusion of Ni^{+2} into MgO, a typical ceramic system [28]. They found that D' was only slightly greater than D, and estimate that $D \simeq D'$ at 2000°C. They also reach the startling conclusion that δ is in the order of microns instead of angstroms. This, they explain, may be caused by impurity segregation at the grain boundary, a nonequilibrium concentration of defects, or space charge effects. They cite similar results in alumina and alkali halides to verify this conclusion [29,30,31]. Additional studies in this area are needed.

The results of grain boundary diffusion analyses are in general agreement with the dislocation model for a grain boundary. If diffusion is more rapid along a boundary, then the diffusion along the length of dislocations should be more rapid. The dislocations form a "pipe" to "short-circuit" the flow. When the concentration gradient causes diffusion along grain boundaries, the grain boundary diffusion is in parallel with the bulk diffusion. When the diffusion is perpendicular to grain boundaries, it is in series with the bulk diffusion.

No suitable model exists to explain why grain boundary diffusion may be faster than bulk diffusion. Li has presented an analysis based on the dislocation core model in which he shows that the diffusion of a tilt boundary should increase rapidly with increased angle until the cores come in contact. It should increase slowly thereafter because the core radius increases in his model [20].

Tilt boundaries should have more rapid diffusion in a direction parallel to the dislocation than in a direction perpendicular to it. This has been observed in metals. The ratio of the excess diffusivity parallel to the dislocations with that perpendicular to the dislocations for a 100 tilt boundary in silver was found to decrease continuously with the angle of tilt up to 45°. The continuous decrease in the ratio over the entire range of angles has led Shewman to criticize models which are characterized as homogeneous isotropic slabs at high angles [32].

Twist boundaries in silver have been found to have a diffusivity only one-tenth that of tilt boundaries [33]. No comparative results are available in ceramic systems. Rapid increases in diffusivity have been found to result when dislocations and grain boundaries are in motion [34]. This may be an important factor in the hot pressing process.

Sintering

Many ceramic products are produced by solid state reaction. In the sintering process, fine powders are pressed and fired. The driving force is surface tension and the mechanisms of sintering may be viscous or plastic flow, evaporation and condensation, volume diffusion, and surface diffusion. Of these only viscous or plastic flow and volume diffusion are effective in causing adequate densification [35]. Kuczynski has shown that when two spheres of radius a are sintered together, a plot of log x/a vs. log t can be used to determine the mechanisms [36]. Here x is the width of the sintered neck where the spheres contact, and t is time. Expressed in the form

$$\frac{x^n}{a^m} = F(T)t$$

where $F(T)$ is a constant for a given temperature, t, the exponents have values depending upon the sintering mechanism:

Mechanism	n	m
Plastic or viscous flow	2	1
Evaporation and condensation	3	1
Volume diffusion	5	2
Surface diffusion	7	4

The effect of grain boundary diffusion may be expected to be extremely important in the sintering process. The porosity of most ceramic materials is extremely important in controlling properties [37] (Fig. 12). To reduce porosity and obtain such desirable properties as high strength, rapid diffusion is called for.

By a more accurate analysis of the shape of particles present during sintering, Johnson and Cutler developed better equations relating shrinkage during sintering to mechanisms of material transport [38]. These equations produced modified Kuczynski exponents for real sintering situations. An analysis of extensive sintering experiments of alumina and magnesia gave the following equation [39]:

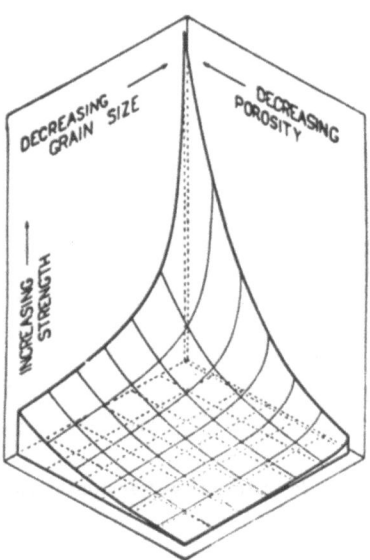

Fig. 12. Effect of grain size and porosity on strength
of polycrystalline ceramic (Knudsen [37]).

$$\frac{L}{L_0 - \Delta L} = 1 - \left(\frac{K^1 D}{T}\right)^{0.31} (t - \Delta t)^{0.31}$$

when L is length, L_0 is initial length, ΔL is a length correction, Δt is a time correction, and

$$K^1 = \frac{50 \delta V \gamma}{7 \pi k a^4}$$

In this expression V is the volume of a vacancy, γ is surface energy, and k is Boltzmann's constant. All other terms are as previously defined.

This equation fits the experimental behavior quite well (Fig. 13). Johnson and Cutler conclude that, for the particle size and temperature range tested, the predominant mechanism of sintering is grain boundary diffusion.

The importance of porosity on properties has led to considerable study of the sintering process in relation to pore removal. Pores on grain boundaries shrink during sintering, but intragranular pores usually do not, even though the length of the diffusion path may be equal. This has been attributed by Paladino and Coble to the rate

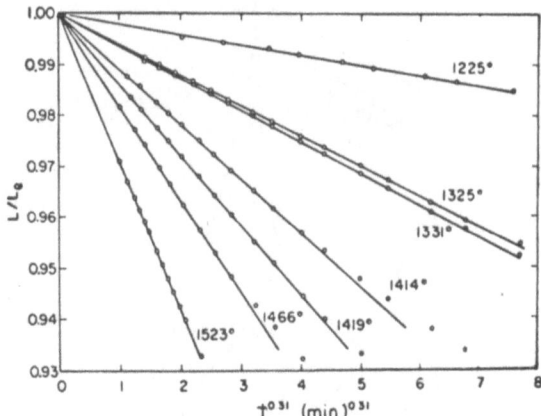

Fig. 13. Sintering isotherms for high purity of alumina (Johnson and Cutler [39]).

of diffusion of oxygen and aluminum ions [40]. When grain boundaries are present, they consider the oxygen ion diffusion rate to be greatly increased in a region around the boundary of 200 A radius. Then the diffusion of oxygen vacancies to the boundary is believed to be 1000 times faster than within the grain. The much slower diffusion of oxygen ions inside the grain prevents the effective removal of intragranular pores.

Ductility

As explained earlier, a grain boundary may serve as a barrier to dislocation motion by its repelling force on dislocations. The fact that certain cubic ionic solids such as MgO, LiF, and NaCl have some ductility as single crystals but no ductility as polycrystals has led to considerable study of the interaction of grain boundaries with dislocations in ceramic crystals. Studies of this type have often led to the conclusion that cracking under these conditions is inevitable, usually by a Stroh-type mechanism [41] (Fig. 14). Stroh suggested that a grain boundary act as a barrier causing dislocations to pile up. When the stress on the leading dislocation becomes great enough to cause the two leading dislocations to coalesce, a crack is formed and fracture follows. This is essentially a stress concentration mechanism. The effective shear stress on the leading dislocation is the applied shear stress, τ, multiplied by the number, n, of dislocations piled up against it. If

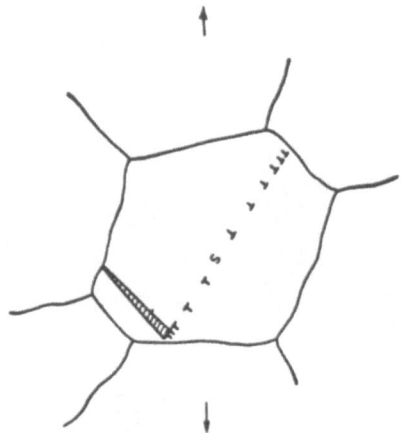

Fig. 14. Stroh model for crack nucleation
(Washburn [42]).

the minimum shear stress required to move a dislocation is τ_i,
the critical stress, τ_c, necessary to form a crack is given by

$$\tau_c - \tau_i = \frac{G}{2\pi\eta\,(1-\nu)}$$

The number of dislocations which can pile up against a boundary of
diameter, λ, will depend on τ_i and λ. Washburn has calculated this
by equating plastic to elastic strain [42]:

$$\frac{nb}{\lambda} = \frac{\tau - \tau_i}{G}$$

$$n = \frac{\lambda\,(\tau - \tau_i)}{G}$$

Then

$$\tau_c = \tau_i + G\left[\frac{b}{2\pi\lambda\,(1-\nu)}\right]^{\frac{1}{2}}$$

Parker has suggested that the Stroh cracking should be ineffective
if λ is less than $1\,\mu$ because insufficient length for stress pile-up
would exist [43].

Some recent studies indicate that grain boundaries in ionic
crystals may not be as embrittling as previously supposed [44,45].
The effect of small-angle tilt and twist boundaries in sodium
chloride has been determined by optical birefringence (Figs. 15, 16,
and 17).

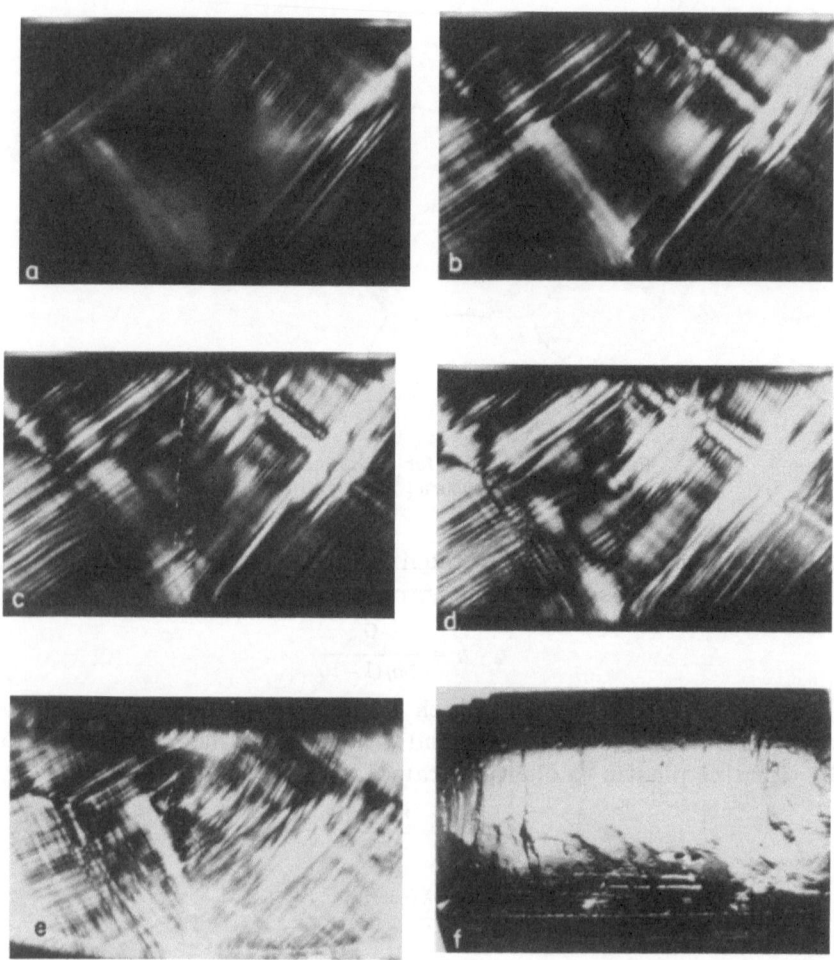

Fig. 15. Birefringence of a 12° tilt bicrystal of sodium chloride in bending. X25 (A) Yield point. (B) Increasing load; slip bands just starting to cross the vertical grain boundary at the upper and lower edges of the bicrystal. (C) Increasing load. Well-formed continuous slip bands exist over most of the boundary. (D) Further increase in load. (E) Further Increase in load.   Entire   boundary   exhibits   continuous   slip. (F) Fracture surface (Long and McGee [45]).

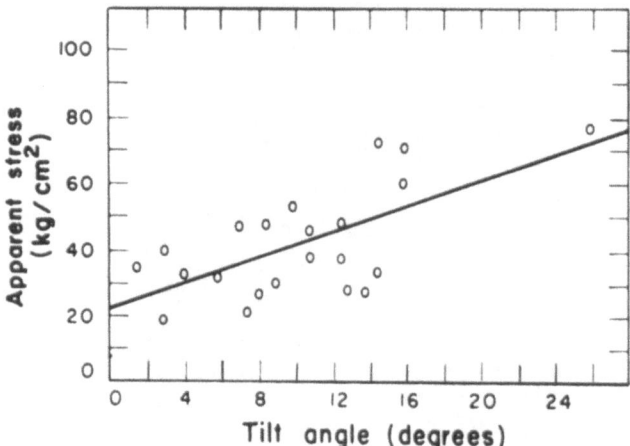

Fig. 16. Apparent stress at which birefringence crossed grain boundaries in sodium chloride as a function of the angle of tilt (Long and McGee [45]).

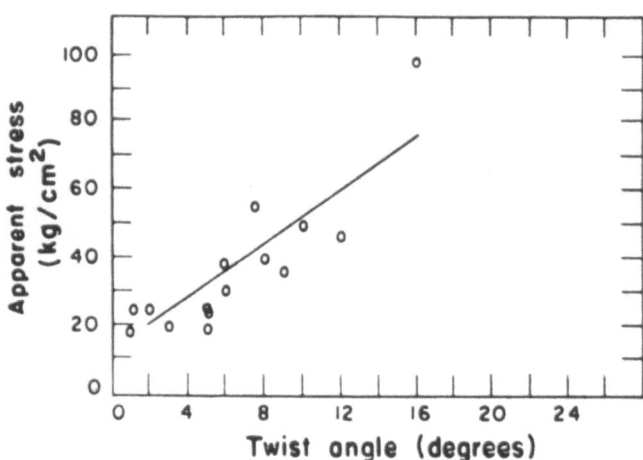

Fig. 17. Apparent stress at which birefringence crosses grain boundaries in sodium chloride as a function of the angle of twist (Long and McGee [45]).

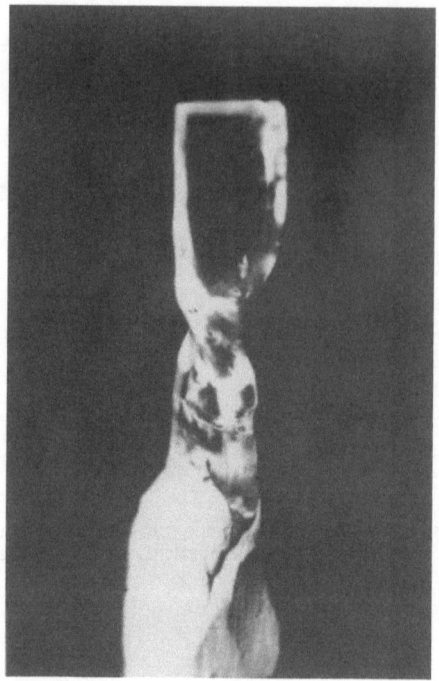

Fig. 18. Sodium chloride bicrystal which was twisted at 600°C.

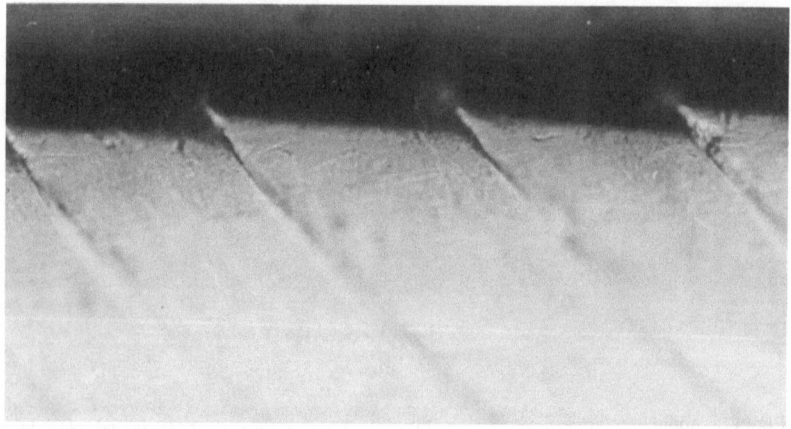

Fig. 19. Glide bands in a sodium chloride bicrystal which was bent at 300°C (×300). Offsets are 0.015 mm spaced 0.1 mm apart.

The 110 slip systems operate in the sodium chloride structure. Cottrell has shown that five slip systems must be active to obtain polycrystalline ductility [46]. Only three independent 110 slip planes are available in NaCl. The activation of 100 planes would greatly improve ductility; the possibility of polycrystalline slip occurring then is the combination of six planes taken five at a time. The increase in ductility which occurs at higher purity and at higher temperatures is partly the result of the activation of 100 slip (Figs. 18 and 19). Whether any possibility of real ductility exists must await further research on the brittle–ductile transition and on grain boundaries.

Strength

Grain size—the presence of grain boundaries—is extremely important to the strength of ceramic materials. This is partly true because of the effect of grain size on the sintering and hot pressing methods of reducing porosity (Fig. 20). It is also true because of the blocking effect of the boundaries (Fig. 21). Because pore removal occurs most effectively through the diffusion of vacancies at grain boundaries, it is necessary to prevent grain growth during heat treatment. This is usually accomplished by the introduction of a very small amount of a second phase, e.g., 0.5–3.0% of MgO added to Al_2O_3 to prevent excessive grain growth during sintering at 1900°C.

Grain boundaries in polycrystalline dense MgO serve to increase the strength above that of single crystals. Stokes and Li found that single-crystal MgO had a tensile strength of 10,000 to 20,000 psi [47]. If no pores were present and the surface was highly polished, polycrystalline MgO had a tensile strength of about 30,000 psi.

Creep

As the temperature is raised, preferential softening or flow often occurs at grain boundaries where impurities may be concentrated. A wide variety of ceramic materials, such as ZrC, MgO, Al_2O_3, BN, and AgCl, have shown creep by grain boundary softening. Ceramic systems are almost identical with metallic systems in their creep behavior at high temperatures. They often exhibit three stages of creep. When grain boundary sliding occurs, it is often jerky because of shearing of grain boundary irregularities lying in the boundary. At the highest temperatures destructive creep occurs by grain boundary parting. This is believed to occur

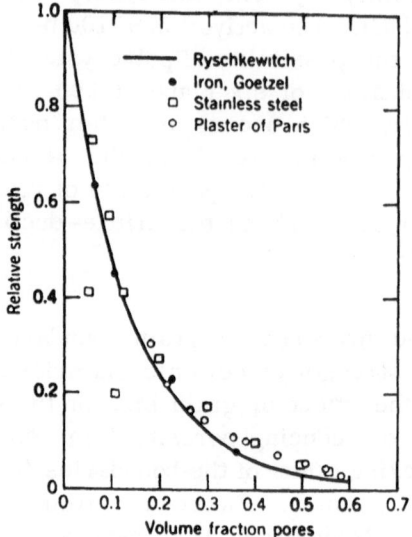

Fig. 20. Effect of porosity on the fracture strength of ceramics (Kingery [35]).

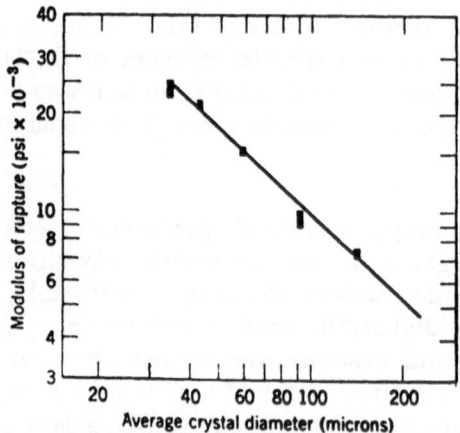

Fig. 21. Effect of grain size on the fracture strength of sintered BeO (Kingery [35]).

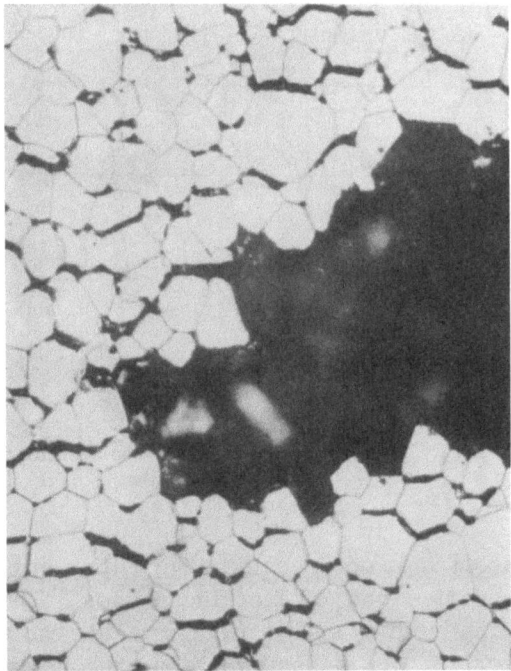

Fig. 22. Grain boundary voids nucleated in the tensile region of boundaries in alumina
(Burke [58]).

by the Nabarro–Herring mechanism where vacancies diffuse to the
tension areas and consolidate at grain boundaries (Fig. 22) [48].
    In a recent study Coble and Guerard analyzed creep in Al_2O_3
[49]. They concluded that the bulk diffusion of aluminum ions was
rate controlling and that oxygen diffused along grain boundaries.
Precipitation of vacancies at grain boundaries caused failure.

Magnetic Properties
    The complex properties of ceramic magnets are also dependent
upon grain boundaries. Domain structures are influenced by the
grain boundaries, pores, and other phases. In the sintering process
the elimination of the pores is dependent on their position in the
grain. Lower porosity almost always brings better magnetic
properties (Fig. 23).
    Many soft ceramic magnets are required to change polarity
rapidly with a minimum of loss. The ideal memory core or high-

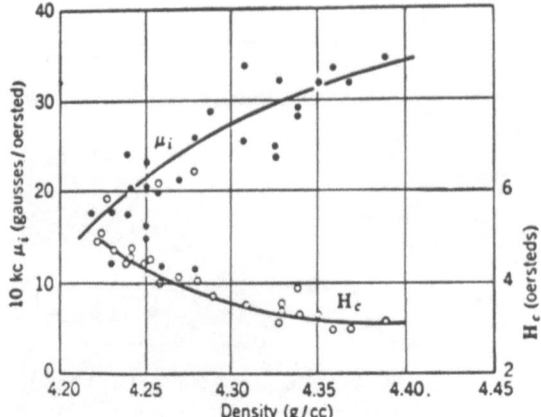

Fig. 23. Effect of density on the initial permeability and coercive force in magnesium
ferrite (Economos [50]).

frequency transformer core would be one with a minimum of pores
of grain boundaries and of blocking phases. The motion of the
Bloch wall is inhibited by such structures. This causes an increase
in the number of Weiss domains and an increase in exchange
energy [50].

In permanent magnets strong crystal anisotropy and alignment
of domains improves the coercive force. By reducing the crystal-
lite size to about 500 A in cobalt ferrite, each grain becomes a
single domain. Excessive reduction in grain size causes a reduction in coercive force because the spins are not effectively aligned
in the disordered region of the grain boundaries (Fig. 24) [51,52].

Polarization

In the manufacture of dielectrics for condensers, the depolarization that occurs at a grain boundary may limit the dielectric constant
that can be obtained. A theoretical analysis of the polarization of
a dielectric was made by Stadelmaier and Derbyshire [53]. They
found that the polarization could be expressed by

$$P = P_0[1 - e^{-\mathcal{N}/\chi^\delta}]$$

where P is polarization, P_0 represents the polarization inside the
grains, χ is the electric susceptibility, and the other terms are as
previously defined.

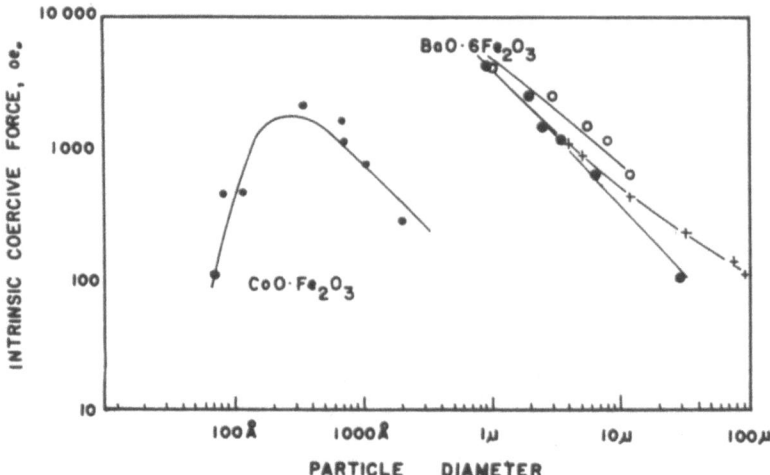

Fig. 24. Effect of particle size on the intrinsic coercive force of cobalt and barium ferrites (Luborsky [52]).

Solution of the equation for barium titanate gave the constants $P_0 = 4 \cdot 10^{-6}$ C/cm^2 and $\delta = 0.9 \cdot 10^{-10}$ M. In this case the polarization was not a factor until the particle size was less than $5\,\mu$. Therefore, grain boundary depolarization was not limiting. The small grain boundary width was attributed to misorientation of boundaries. In the derivation all boundaries were assumed to be parallel to the applied field.

Conductivity

The electrical conductivity of dielectrics may be increased by grain boundary conduction because of the enhanced diffusion and because of impurity separation. Tien has studied lime–stabilized zirconia and found that the conductivity was a function of the grain size (Fig. 25) [54]. If the following expression is used for conductivity,

$$\sigma = \left[1 - \left(\frac{2\delta}{\lambda}\right)\right]\sigma_{\text{grain}} + \left(\frac{2\delta}{\lambda}\right)\sigma_{\text{grain boundary}}$$

the grain boundary conductivity at 600°C is found to be 200 times greater than the intragranular conductivity.

Transparency

Pure single crystals are often transparent. Polycrystalline ceramic materials usually are not. The transparency is a function

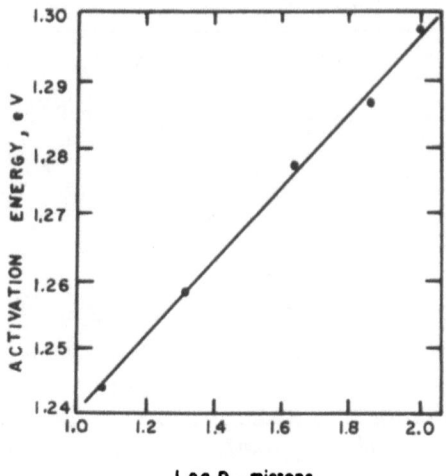

Fig. 25. Activation energy for electrical conduction as a function of the grain size of lime stabilized zirconia (Tien [54]).

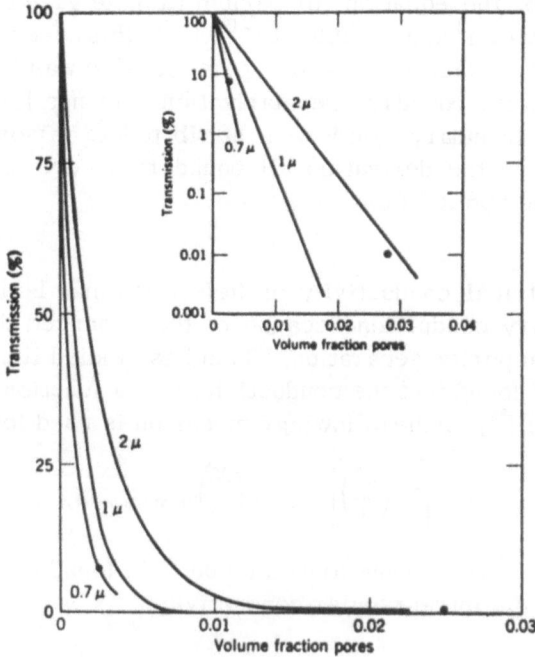

Fig. 26. Transmission of polycrystalline alumina for visible wave lengths as a function of pore fraction and pore size (Kingery [35]).

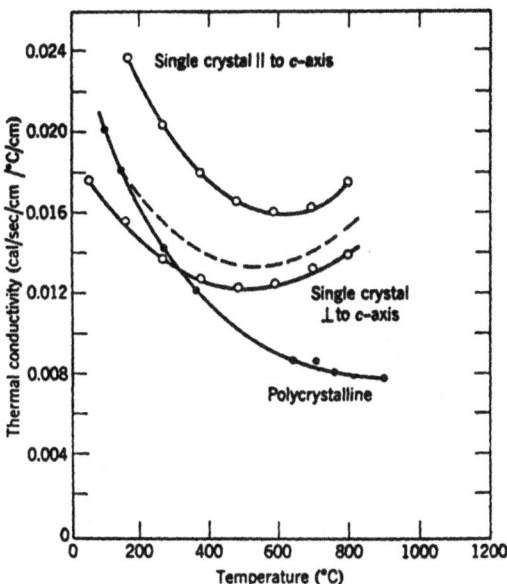

Fig. 27. Thermal conductivity of single crystal and polycrystalline titania (Kingery [35]).

of the index of refraction of the specimen. If pores are present, transparency will not exist. The degree of transmission is sensitive to extremely small quantities of voids or other scattering centers (Fig. 26) [55].

Radiant heat transfer is, of course, a function of transparency at high temperatures. The usual decrease in thermal diffusivity as temperature is increased is reversed when radiant heat transfer becomes effective. The importance of grain boundaries in decreasing the mean free path by phonon scattering is shown when single crystal and polycrystalline specimens are compared (Fig. 27) [36].

Thermal Emission

According to Ryshkewitch the use of thoria containing 1% ceria as a light source is controlled by the grain size [56]. The actual source of light emission is in the grain boundary. The optimum grain size was found to be dependent on the index of refraction and increased with the index. The optimum grain size for thoria was 2–4 μ.

IV. SUMMARY

Some of the theoretical aspects of grain boundaries in ceramic materials have been reviewed. The presence of grain boundaries may control such properties as ductility, internal friction, strength, creep, magnetic behavior, polarization, conductivity, transparency, and thermal emission.

V. ACKNOWLEDGMENTS

The support of the U. S. Army Research Office (Durham) is gratefully acknowledged.

REFERENCES

1. S. Amelinckx and W. Dekeyser, "The Structure and Properties of Grain Boundaries," in Solid State Physics, Advances in Research and Applications, Vol. 8, edited by F. Seitz and D. Turnbull (Academic Press, London, 1959), pp. 325–499.
2. J. M. Burgers, "Geometrical Considerations Concerning the Structural Irregularities to Be Assumed in a Crystal," Proc. Phys. Soc. (London) A52(1):23–33 (1040).
3. W. L. Bragg, Discussion of reference 2, Proc. Phys. Soc. (London) A52(1):54–55 (1940).
4. W. Shockley and W. T. Read, "Quantitative Predictions from Dislocation Models of Grain Boundaries," Phys. Rev. 75(4):692 (1949).
5. J. H. Van der Merwe, "On the Stresses Associated with Inter-Crystalline Boundaries," Proc. Phys. Soc. (London) A63(6):616–637 (1950).
6. W. T. Read, Jr., Dislocations in Crystals (McGraw-Hill Book Company, Inc., New York, 1953).
7. M. L. Kronberg, "Plastic Deformation of Single Crystals of Sapphire: Basal Slip and Twinning," Acta Met. 5(9):507–524 (1957).
8. J. Smit and H. P. Wign, Ferrites (John Wiley and Sons, New York, 1959).
9. S. Amelinckx and P. Delavignette, "Dislocations in Layer Structures," Chapt. 8, in Electron Microscopy and Strength of Crystals, edited by G. Thomas and J. Washburn, (Interscience Publishers, New York, 1963), pp. 441–513.
10. A. W. Sleeswyk, "The Tilt Angle" (Letter to the Editor), Acta. Met. 11(12):1192–3 (1963).
11. H. G. Van Buren, "Imperfections in Crystals," 2nd edition (North-Holland Publishing Company, Amsterdam, 1961).
12. Doris Kuhlmann-Wilsdorf, "Some Theoretical Considerations on the Geometry of Low-Angle Dislocation Boundaries," J. Appl. Phys. 33(2):648–654 (1962).
13. James C. M. Li, "The Interaction of Parallel Edge Dislocations with a Simple Tilt Dislocation Wall," Acta Met. 8(5):296–311 (1960).
14. W. T. Read and W. Shockley, "Dislocation Models of Grain Boundaries," Phys. Rev. 78(3):275–289 (1950).
15. D. McLean, "Grain Boundaries in Metals," (Oxford University Press, London,1957).
16. Sir G. T. Beilby, Aggregation and Flow of Solids (Macmillan, London, 1921).
17. N. F. Mott, "Slip at Grain Boundaries and Grain Growth of Metals," Proc. Phys. Soc. (London) 60(4):391–394 (1948).
18. W. Shockley, "Dislocation Model of Grain Boundaries," in L'état Solide (Report of 9th International Solvay Conference, Brussels, 1951).
19. B. Chalmers, "Structure of Crystal Boundaries," Progr. in Metal Phys. 3:293–319, 1952.
20. James C. M. Li, "High Angle Tilt Boundary—A Dislocation Core Model," J. Appl. Phys. 32(3):525–541 (1961).

21. T.S. Ke, "A Grain Boundary Model and the Mechanism of Viscous Intercrystalline Slip," J. Appl. Phys. 20(3):274—280 (1949).
22. J.H. Van der Merwe, "Crystal Interfaces. Part I. Semi—Infinite Crystals," J. Appl. Phys. 34(1):117—122 (1963).
23. J.H. Van der Merwe, "Crystal Interfaces. Part II. Finite Overgrowths," J. Appl. Phys. 34(1):123—127 (1963).
24. Paul G. Shewman, Diffusion in Solids (McGraw-Hill Book Co., New York, 1963).
25. A.D. LeClaire, "The Analysis of Grain Boundary Diffusion Measurements," Brit. J. Appl. Phys. 14(6):351—356 (1963).
26. R.T.P. Whipple, "Concentration Contours in Grain Boundary Diffusion," Phil. Mag. 45(371):1225—1236 (1954).
27. J.C. Fisher, "Calculation of Diffusion Penetration Curves for Surface and Grain Boundary Diffusion," J. Appl. Phys. 22(1):74—77 (1951).
28. B.J. Wuensch and T. Vasilos, "Grain Boundary Diffusion in MgO," J. Am. Ceram. Soc. 47:63—68 (1963).
29. J.F. Laurent and Jacques Bénard, "Determination de l'autodiffusion das le chlorure de sodium mono et polycrystalin" ("Determination of Self-Diffusion in Mono- and Polycrystalline Sodium Chloride"), Compt. Rend. 241(8):1204—1207 (1955).
30. J.F. Laurent and Jacques Bénard, "Autodiffusion des ions dans les halogenures alcalins polycrystallins" ("Self-Diffusion of Ions is Polycrystalline Alkali Halides"), Phys. and Chem. Solids 7(2—3):218—27 (1958).
31. L.W. Barr, I.M. Hoodless, J.A. Morrison, and R. Rudham, "Effects of Gross Imperfections on Chloride—Ion Diffusion in Crystals of Sodium Chloride and Potassium Chloride," Trans. Faraday Soc. 56(449):697—708 (1960).
32. Paul G. Shewman, Diffusion in Solids (McGraw-Hill Book Co., New York, 1963), p. 175.
33. G. Love and P.G. Shewman, "Self—Diffusivity of Silver in Twist Boundaries," Acta Met. 11(8):899—906 (1963).
34. A.L. Ruoff and R.W. Balluffi, "Strain-Enhanced Diffusion in Metals. II. Dislocation and Grain Boundary Short-Circuiting Models," J. Appl. Phys. 34(7):1848—1853 (1963).
35. W.D. Kingery, Introduction to Ceramics, (John Wiley and Sons, New York, 1960).
36. H. Ichinose and G.C. Kuczynski, "Role of Grain Boundaries in Sintering," Acta Met. 10(3):209—213 (1962).
37. F.P. Knudsen, "Dependence of Mechanical Strength of Brittle Polycrystalline Specimens on Porosity and Grain Size," J. Am. Ceram. Soc. 42(8):376—387 (1959).
38. D. Lynn Johnson and Ivan B. Cutler, "Diffusion Sintering: I. Initial Stage Sintering Models and Their Application to Shrinkage of Powder Compacts," J. Am. Ceram. Soc. 46(11):541—545 (1963).
39. D. Lynn Johnson and Ivan B. Cutler, "Diffusion Sintering: II. Initial Sintering Kinetics of Alumina," J. Am. Ceram. Soc. 46(11):545—550 (1963).
40. A.E. Paladino and R.L. Coble, "Effect of Grain Boundaries on Diffusion—Controlled Processes in Aluminum Oxide," J. Am. Ceram. Soc. 46(3):133—136 (1963).
41. A.N. Stroh, "The Formation of Cracks as a Result of Plastic Flow," Proc. Royal Soc. (London) A223(1154):404—414 (1955).
42. Jack Washburn, "Mechanism of Fracture," Chapt. 6, in Mechanical Behavior of Materials at Elevated Temperatures, edited by John E. Dorn (McGraw-Hill Book Co., New York, 1961), pp. 108—128.
43. Earl R. Parker, "Ductility of Magnesium Oxide," in Mechanical Properties of Engineering Ceramics, edited by W. Wurth Kriegel and Hayne Palmour III, (Interscience Publishers, New York, 1961), pp. 65—87.
44. S. Feuerstein and E.R. Parker, "The Effect of Grain Boundaries on the Mechanical Properties of Ionic Crystals," 5th Technical Report Issue No. 5, Series No. 150, Materials Research Laboratory, Institute of Engineering Research, University of California (Berkeley), 1962.
45. Stanley A. Long and Thomas D. McGee, "Effect of Grain Boundaries on Plastic Deformation of Sodium Chloride," J. Am. Ceram. Soc. 46(12):583—587 (1963).
46. A.H. Cottrell, Dislocations and Plastic Flow in Crystals (Oxford University Press, London, 1953).
47. R.J. Stokes and C.H. Li, "Dislocations and the Tensile Strength of Magnesium Oxide," J. Am. Ceram. Soc. 46(9):423—434 (1963).
48. R.C. Folweiler, "Creep Behavior of Pore-Free Polycrystalline Aluminum Oxide," J. Appl. Phys. 32(5):773—778 (1961).

49. R. L. Coble and Y. H. Guerard, "Creep of Polycrystalline Aluminum Oxide," J. Am. Ceram. Soc. 46(7):353–354 (1963).
50. George Economos, "Effect of Microstructure on the Electrical and Magnetic Properties of Ceramics," Chapt. 21 in Ceramic Fabrication Processes, edited by M. D. Kingery (John Wiley and Sons, New York, 1960), pp. 201–213.
51. W. J. Schuele and V. D. Deetscreek, "Fine Particle Ferrite," in Ultrafine Particles, edited by W. E. Kuhn (John Wiley and Sons, New York, 1963), pp. 218–235.
52. Fred E. Luborsky, "The Application of Ultrafine Particles to the Fabrication of Permanent Magnets," in Ultrafine Particles, edited W. E. Kuhn (John Wiley and Sons, New York, 1963), pp. 488–513.
53. H. H. Stadelmaier and S. W. Derbyshire, "Relation between Electrical Properties and Microstructure of Barium Titanate," in Materials Science Research, Vol. 1, edited by H. H. Stadelmaier and W. W. Austin (Plenum Press, New York, 1963), pp. 57–65.
54. T. Y. Tien, "Grain Boundary Conductivity of $Zr_{0.84} Ca_{0.16} O_{1.84}$ Ceramics," J. Appl. Phys. 35(1):122–124 (1964).
55. D. W. Lee and W. D. Kingery, "Radiation Energy Transfer and Thermal Conductivity of Ceramic Oxides," J. Am. Ceram. Soc. 43(11):594–607 (1960).
56. Eugene Ryshkewitch, Oxide Ceramics (Academic Press, New York, 1960).
57. S. Amelinckx, "Dislocations in Ionic Crystals: I. Geometry of Dislocations" Chapt. 2 in Mechanical Properties of Engineering Ceramics, edited by W. Wurth Kriegel and Hayne Palmour III, (Interscience Publishers, New York, 1961), pp. 9–33.
58. J. E. Burke, "Grain Boundary Effects in Ceramics," in Materials Science Research, Vol. 1, edited by H. H. Stadelmaier and W. W. Austin (Plenum Press, New York, 1963), pp. 69–87.

Defect Structure and Electrical Properties of Some Refractory Metal Oxides

N. M. Tallan*

Aerospace Research Laboratories, Wright-Patterson Air Force Base, Ohio

R. W. Vest [†]

Systems Research Laboratories, Inc., Dayton, Ohio

and H. C. Graham [‡]

Aerospace Research Laboratories, Wright-Patterson Air Force Base, Ohio

A brief description of the equilibrium thermodynamic approach to the characterization of defect concentrations in refractory metal oxides as a function of temperature, oxygen partial pressure, and impurity content is given. Techniques for the determination of ionic and electronic transport numbers by a blocking electrode polarization measurement and for the measurement of conductivity, thermoelectric power, and weight change are reviewed. The application of this approach and these measurements to the determination of the extent of deviation from stoichiometry, the nature and ionization state of the defects which predominate, and the mechanism of charge transport are illustrated by detailed consideration of several specific examples.

Measurements on calcium-stabilized zirconia are cited as an example of an oxide in which the defect structure and conductivity are controlled by aliovalent foreign ions. This material is intrinsically a pure oxygen-ion conductor down to very low temperatures, but some electronic conductivity may be observed in the presence of certain impurities. Measurements on pure monoclinic zirconia at 1000°C will be presented as an example of a transition metal oxide whose transport properties are essentially determined by its deviations from stoichiometry. It is shown, for example, that, at 1000°C and high oxygen pressures, fully-ionized metal vacancies predominate and charge transport occurs by low mobility holes.

I. DEFECT STRUCTURE OF THE OXIDES

The importance of the defect structure in a material in directly determining its transport properties and influencing the rates at which many processes occur has long been recognized by metallur-

*Supervisor, Ceramic Research.
†Senior Scientist.
‡Research Physicist.

gists.   The role of these defects in ceramic materials is now re-
ceiving increasing attention, but these studies have been hampered
by inherent complexities that generally do not arise in the study
of metals.   In addition to the obvious presence of a greater variety
of defects, both cation and anion vacancies, interstitials, and impurities, one has to consider carefully deviations from stoichiometry
induced by oxidation or reduction of the material and the ionization
steps producing the various charged defect species and finite concentrations of electrons and holes. Before considering experimental
techniques for determining the predominant defects in a metal oxide
and specific examples of the application of these techniques to
several oxides, it would be well to consider, in order, the defects
generated thermally in a pure, stoichiometric compound, the modifications induced by deviations from stoichiometry, and finally the
effect of impurities.

Thermally-Generated Defects in Pure Stoichiometric Oxides

In the case of a pure, stoichiometric oxide, MO, defects can be
generated thermally by several familiar processes: the formation of
equivalent concentrations of cation and anion vacancies (Schottky–
Wagner disorder), the formation of equivalent concentrations of
vacancies and interstitials on either the cation or anion sublattice
(Frenkel disorder), or the apparently very rare interchange of
positions between cations and anions (antistructural disorder).
Fortunately, the concentrations of the defects generated by these
processes are not independent of each other.   This can be demonstrated by considering the defects generated thermally by Schottky–
Wagner disorder, which, following the notation of Kroger and
Vink [1], can be described by the pseudoreaction

$$(1 - \delta) \, MO \rightleftharpoons M_{(1 - \delta)} \, O_{(1 - \delta)} + \delta V_M + \delta V_O - E_S \qquad (1.1)$$

where the metals and oxygens on their normal sites are represented by M and O without any subscript, an imperfection is represented by a symbol indicating its nature (i.e., an atom or a vacancy),
and a subscript indicating its position; δ is the concentration of the
imperfection (assumed to be very small), and E_S is the energy
involved.   In writing the reaction this way, the vacancies generated
have been assumed to be independent of each other. The associa-
tion of these defects into pairs, which can be represented by the
reaction

$$V_M + V_O \rightleftharpoons (V_M \; V_O) + E_B \qquad (1.2)$$

where E_B is the binding energy, can be included in a complete description when necessary, but will be excluded here.

Each of the neutral vacancies generated in reaction (1.1) will in turn generate electrons and holes by a series of ionization steps which can also be written as pseudoreactions:

$$V_O \rightleftharpoons V_O^{\cdot} + \ominus - E_1 \qquad (1.3)$$

$$V_O^{\cdot} \rightleftharpoons V_O^{\cdot\cdot} + \ominus - E_2 \qquad (1.4)$$

and so on through as many ionization steps as are energetically realistic for the anion vacancy, and

$$V_M \rightleftharpoons V_M{}' + \oplus - E_3 \qquad (1.5)$$

$$V_M{}' \rightleftharpoons V_M{}'' + \oplus - E_4 \qquad (1.6)$$

and so on, where $\ominus$ represents an electron, $\oplus$ a hole, the dot a positive charge, the prime a negative charge, and E_1 to E_4 are the pertinent ionization energies.

In addition to the electrons and holes generated by ionization of these defects, electrons and holes may be generated by intrinsic processes, such as the promotion of an electron from the valence band to the conduction band in a material for which semiconductor band theory applies or the transfer of an electron from one cation to another in a transition-metal oxide. This intrinsic generation of electronic carriers can be represented by the reaction

$$\text{null} \rightleftharpoons \ominus + \oplus - E_i \qquad (1.7)$$

The concentrations of the defects generated by reactions (1.1) and (1.3)–(1.7) are related, under equilibrium conditions, through the application of the law of mass action to each reaction as follows:

$$[V_M] \, [V_O] = K_S = C_S \exp{(-\Delta H_S / RT)} \qquad (1.8)$$

$$n \cdot p = K_i = C_i \exp{(-\Delta H_i / RT)} \qquad (1.9)$$

$$\frac{n[V_O^{\cdot}]}{[V_O]} = K_1 = C_1 \exp{(-\Delta H_1 / RT)} \qquad (1.10)$$

$$\frac{n[V_O^{\cdot\cdot}]}{[V_O^{\cdot}]} = K_2 = C_2 \exp{(-\Delta H_2 / RT)} \qquad (1.11)$$

$$\frac{p[V_M{}']}{[V_M]} = K_3 = C_3 \exp{(-\Delta H_3 / RT)} \qquad (1.12)$$

$$\frac{p[V_M{}'']}{[V_M{}']} = K_4 = C_4 \exp\left(-\Delta H_4/RT\right) \tag{1.13}$$

where the K's are equilibrium constants and the C's are constants containing the entropy change. In addition to these relations, in order to maintain stoichiometry and electroneutrality, the relations

$$[V_M] + [V_M{}'] + [V_M{}''] = [V_O] + [V_O{}^{\cdot}] + [V_O{}^{\cdot}] \tag{1.14}$$

and

$$n + [V_M{}'] + 2[V_M{}''] = p + [V_O{}^{\cdot}] + 2[V_O{}^{\cdot}] \tag{1.15}$$

must hold. The unknown defect concentrations are uniquely related by equations (1.8)–(1.15) and can be computed if the equilibrium constants involved are known.

Since the defect concentrations are also explicitly related to the energies required for their formation by equations (1.8)–(1.15), the measurement of any property of a material which is directly proportional to the defect concentrations can yield values for the energies of formation of those defects. For example, if the electrical conductivity is measured over a sufficiently wide range of temperature that regions governed by (1) the energy for carrier motion alone, (2) the energy for motion and ionization, and (3) the energy for motion, ionization, and formation are all observed, then these energies can be separated. However, the identification of the predominant defects present solely by means of the observed energies is clearly difficult in most practical cases.

From the ionization reactions (1.3)–(1.6), it is clear that the neutral V_O and the partially ionized $V_O{}^{\cdot}$ are electron donors and that similarly V_M and $V_M{}'$ are electron acceptors. In the simple case of a stoichiometric metal oxide MO in which the donor and acceptor levels are symmetrically placed within the band gap, i.e., when $E_1 = E_3$ and $E_2 = E_4$, then the numbers of electrons and holes generated at any temperature will be equal; then, if the recombination rate is high, the donor and acceptor levels will be essentially self-compensating and very few electrons or holes will be available for conduction. However, if the donor and acceptor levels are asymmetric, then even the pure stoichiometric oxide MO will be either n-type or p-type unless the number of electrons and holes generated by an intrinsic process represented by reaction (1.7) exceeds the number generated by the ionization reactions (1.3)–(1.6). For example, if $E_1 < E_3$ and $E_2 < E_4$, then after the acceptor levels are compensated, there will be remaining electrons available for conduction; the oxide will be n-type until a transition temperature is reached above which reaction (1.7) predominates. In the case of a more general oxide composition, M_nO_m, where $n \neq m$, it would be fortuitous indeed if there were complete compensation of the donor and acceptor levels; in general the pure stoichiometric oxide will be either n-type or p-type until an intrinsic process predominates.

Finally, while it was stated earlier that defects could be generated thermally by Schottky—Wagner, Frenkel, or antistructural disorder, generally only one process need be considered because ordinarily the energies required for the different processes will be quite different for any given material, and the one requiring the lowest energy will predominate. However, there can in principle be cases in which the energies for two or more processes will be comparable, and then the reaction representing each must be included in the set of simultaneous equations to be solved. While in this event the relations between the defect concentrations represented by each mass action equation must still hold, if a given defect is generated simultaneously by two or more processes, its concentration will be less than the sum of the concentrations predicted from each individual process.

Modifications Induced by Deviations from Stoichiometry

The tendency for an oxide to vary in composition with variations in the oxygen pressure in the surrounding vapor phase, or alternatively with the metal pressure since the two pressures are related by the energy of formation of the metal oxide, can be represented for the oxide MO_2 by reactions of the form

$$O_O \rightleftharpoons \frac{1}{2}(O_2)_g + V_O - E_{red.} \tag{1.16}$$

$$M_M + 2O_O \rightleftharpoons (O_2)_g + M_i - E_{red.} \tag{1.17}$$

depending on whether the reduction results in oxygen vacancies or metal interstitials, and

$$(O_2)_g \rightleftharpoons 2O_O + V_M - E_{ox.} \tag{1.18}$$

$$\frac{1}{2}(O_2)_g \rightleftharpoons O_i - E_{ox.} \tag{1.19}$$

depending on whether the oxidation results in metal vacancies or oxygen interstitials. Under equilibrium conditions, the law of mass action can be applied to reactions (1.16)—(1.19) to obtain relations between the concentrations of the neutral defects formed as a result of deviation from stoichiometry and the oxygen pressure. In the specific case of an oxide MO_2 in which thermal generation of defects is predominately by Schottky–Wagner disorder, for example, reactions (1.16) and (1.18) will predominate since the energetics of interstitial formation have been assumed to be unfavorable in neglecting Frenkel disorder. In this case, application of the mass action law yields the relations

$$[V_O] = K_{red.}\ p_{O_2}^{-1/2} \tag{1.20}$$

$$[V_M] = K_{ox.}\ p_{O_2} \tag{1.21}$$

While these relations indicate the simple pressure dependences of the neutral vacancy concentrations, the pressure dependences of the charged defect concentrations can be obtained only by considering the overall problem in detail; intuitive assumptions of the degree of ionization of the defects formed and the oxygen pressure dependence of the electronic charge carrier concentrations are often deceptive. This can best be illustrated by considering the detailed solution to the relatively simple problem already outlined, the oxide MO_2 with preferential vacancy formation by Schottky–Wagner disorder and deviation from stoichiometry. The complete set of relations between the defects is

$$[V_M]\,[V_O]^2 = K_s \tag{1.22}$$

$$[V_M] = K_{ox.}\ p_{O_2} \tag{1.23}$$

$$\frac{n[V_O^{\cdot}]}{[V_O]} = K_1 \tag{1.24}$$

$$\frac{n[V_O^{\cdot\cdot}]}{[V_O^{\cdot}]} = K_2 \tag{1.25}$$

$$\frac{p[V_M']}{[V_M]} = K_3 \tag{1.26}$$

$$\frac{p[V_M'']}{[V_M']} = K_4 \tag{1.27}$$

$$\frac{p[V_M''']}{[V_M'']} = K_5 \tag{1.28}$$

$$\frac{p[V_M'''']}{[V_M''']} = K_6 \tag{1.29}$$

$$n \cdot p = K_i \tag{1.30}$$

$$n + [V_M'] + 2[V_M''] + 3[V_M'''] + 4[V_M''''] = p + [V_O^{\cdot}] + 2[V_O^{\cdot\cdot}] \tag{1.31}$$

if the metal is quadrivalent. Following the procedure outlined by Kroger and Vink [1], one can solve these ten equations in ten unknown defect concentrations easily, using the Brouwer [2] assumption that one quantity predominates on each side of equation (1.31) in any given oxygen pressure range, to simplify the electroneutrality condition. In this case there would be fifteen different simplified neutrality conditions; solving equations (1.22)—(1.30) with each of these in turn, fifteen analytical solutions for the pressure dependence of the defect concentrations would be obtained, each of which would be valid in the range of oxygen pressures in which the simplified neutrality condition used is valid. Having these general analytical solutions, all of the possible pressure dependences that could be obtained for each of the defect concentrations and the simplified neutrality conditions that would yield these pressure dependences could be tabulated. With the help of a table like this, the experimentally determined pressure dependence of any given defect concentration could be used to reduce to a small number the possible sets of analytic solutions for the defects present and, in some cases, even to determine uniquely the pertinent solution for that temperature and oxygen pressure range.

Since, at a given oxygen pressure, the defect concentrations are proportional to the equilibrium constants appearing in equations (1.22)–(1.30), in general not all of the potentially possible simplified neutrality conditions will be valid for any given set of equilibrium constants. For example, with the entirely arbitrary set of constants, $K_s = 10^{46}$ cm^{-9}, $K_1 = K_3 = 10^{20}$ cm^{-3}, $K_2 = K_4 = 10^{17}$ cm^{-3}, $K_5 = 10^{15}$ cm^{-3}, $K_6 = 10^{13}$ cm^{-3}, and $K_i = 10^{33}$ cm^{-6}, only five of the fifteen possible simplified neutrality conditions would be valid; the pressure dependences of the defect concentrations could be represented graphically by considering only those five pressure ranges for which one of these conditions is valid. The graphic solution for this specific set of constants, chosen only so that Schottky–Wagner disorder would be energetically somewhat more favorable than intrinsic electronic carrier formation and that the energy for the fourth ionization step of the metal vacancy would correspond roughly to

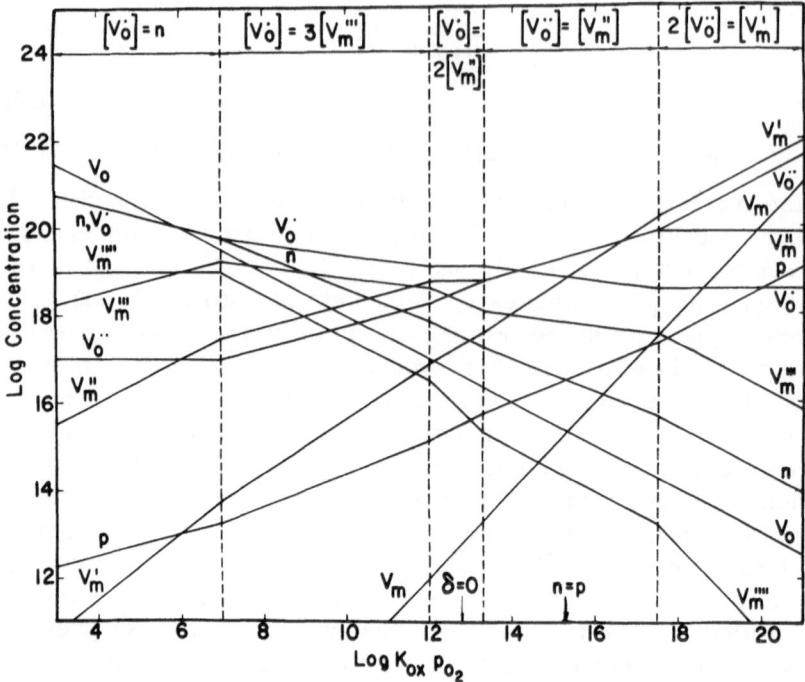

Fig. 1. Concentrations of the imperfections in an oxide MO_2 with Schottky disorder.

the center of the band gap, is presented in Fig. 1. While other
graphic solutions involving other simplified neutrality conditions
could be obtained by selecting other sets of equilibrium constants,
the significant points to be noted can be illustrated by this simple
case. It is immediately apparent in Fig. 1 that under these conditions: (a) although the total oxygen vacancy concentration is high at
low oxygen pressures, it is the singly-ionized oxygen vacancy that
predominates; the fully ionized oxygen vacancy does not predominate until higher oxygen pressures (in fact, in this case until the high
oxygen pressure side of the stoichiometric point denoted by $\delta = 0$ is
reached), (b) with the specific set of equilibrium constants selected,
the fully-ionized metal vacancy concentration never predominates
in the electroneutrality condition, (c) although the total metal
vacancy concentration is high at high oxygen pressures, it is the
singly-ionized metal vacancy that predominates (the fully-ionized
metal vacancy predominates over the other ionization states of the
metal vacancy at low oxygen pressures), (d) with this specific set of
equilibrium constants the electron "hole" concentration at high

oxygen pressures increases with oxygen pressure at the same rate as the fully-ionized oxygen vacancy concentration, and therefore it will never predominate in the electroneutrality condition, (e) in this specific case, on the high oxygen pressure side of the stoichiometric point, $\delta = 0$, the total oxygen vacancy concentration is still increasing with increasing oxygen pressure, (f) as a result of the oxide composition MO_2 and the resultant asymmetry of the donor and acceptor levels in the band gap, the oxygen pressure at which the electron and hole concentrations are equal (the point at which $n = p$) does not coincide with the oxygen pressure at which the material is stoichiometric.

While the experimentally-determined oxygen pressure dependence of properties proportional to the defect concentrations can be used to great advantage to determine the nature of the defects present and their ionization states, it is clear from this example that the technique must be applied with caution. Although the equilibrium constants chosen for Fig. 1 were arbitrary, and it will be shown later that experimental results for ZrO_2 indicate that the constants for that oxide do not have the relative magnitudes selected, it is apparent that to obtain a valid understanding of the oxygen pressure dependences of the defect concentrations the most complete set of defect-forming reactions that can be considered, within obvious practical limitations, should be employed. A complete tabulation of the pressure dependences which may be obtained for each defect concentration would in general be most helpful.

The Effect of Impurities

Impurities may be incorporated into an oxide on cation, anion, or interstitial sites; but, in general, the ionization potentials for the impurity levels will not be the same as those for the constituents of the host material. In principle, the effect of an impurity on the concentrations of the defects normally present can be determined by a modification of the analytical procedure already outlined in the preceding section. In addition to the reactions for defects generated thermally and by oxidation–reduction, the charged impurity concentrations would be added to the neutrality condition, ionization reactions would be included for the formation of each of the possible charged impurity species, an equation would be included to describe the fact that the sum of the concentrations of the neutral and charged impurity species must equal the total impurity content, and the resultant set of equations would be solved simultaneously.

Thus, for the case of a foreign atom with two possible donor levels, such as calcium, substituted for a quadrivalent metal, such as zirconium, in the oxide MO_2 with predominantly Schottky—Wagner disorder, the equations to be considered would be

$$[V_M][V_O]^2 = K_s \tag{1.32}$$

$$[V_M] = K_{ox}. p_{O_2} \tag{1.33}$$

$$\frac{n[V_O^{\cdot}]}{[V_O]} = K_1 \tag{1.34}$$

$$\frac{n[V_O^{\cdot\cdot}]}{[V_O^{\cdot}]} = K_2 \tag{1.35}$$

$$\frac{p[V_M^{'}]}{[V_M]} = K_3 \tag{1.36}$$

$$\frac{p[V_M^{''}]}{[V_M^{'}]} = K_4 \tag{1.37}$$

$$\frac{p[V_M^{'''}]}{[V_M^{''}]} = K_5 \tag{1.38}$$

$$\frac{p[V_M^{''''}]}{[V_M^{'''}]} = K_6 \tag{1.39}$$

$$\frac{p[F_M^{'}]}{[F_M]} = K_7 \tag{1.40}$$

$$\frac{p[F_M^{''}]}{[F_M^{'}]} = K_8 \tag{1.41}$$

$$n \cdot p = K_i \tag{1.42}$$

$$[F_M] + [F_M^{'}] + [F_M^{''}] = [F_M]_{total} \tag{1.43}$$

$$n + [V_M^{'}] + 2[V_M^{''}] + 3[V_M^{'''}] + 4[V_M^{''''}] + [F_M^{'}] + 2[F_M^{''}]$$
$$= p + [V_O^{\cdot}] + 2[V_O^{\cdot\cdot}] \tag{1.44}$$

While the analytical solution of this problem in its entirety would be considerably more laborious than the problem outlined in the preceding section, its solution would be essential for a clear understanding of the pressure dependences of the defect concentrations in this material with a small concentration of the impurity. In general, the impurities present strongly influence the defect concentrations in the oxygen pressure range where the material is nearly stoichiometric; at oxygen pressures sufficiently high or low that oxidation or reduction of the material dominates the defect structure, the impurities play only a very minor role. When the impurity concentration is high, as it is particularly in a material like calcium-stabilized zirconia where the calcium content is generally 15%, the pressure ranges where oxidation or reduction of the material itself would predominate may be so extreme that, within the oxygen pressure range that can be conveniently studied, only the pressure dependences of the impurity-controlled defects may be observed.

Several kinds of defect-controlled behavior may be encountered. For example, in the case of substitutional impurities, if at a given oxygen pressure the predominant charged impurity species does not have the same charge as the normal constituents of the material, then either vacancies will be generated or some of the normal constituents will change their effective charge, whichever is the energetically more favorable process, in order to maintain electroneutrality. Thus, metal vacancies are generated when divalent metal impurities are present in a monovalent nontransition metal compound like an alkali halide, but the incorporation of a monovalent metal impurity in a divalent transition metal compound like NiO may be electrically compensated by the formation of an equivalent number of trivalent nickel ions. When the impurities present can have several different effective charges depending on the oxygen pressure in the surroundings, then clearly the pressure dependences of the defect concentrations may be quite complex.

II. EXPERIMENTAL TECHNIQUES

Transference Numbers

When a metal oxide is placed in an electric field, there are four different charged species which can possibly move. These are metal ions, oxygen ions, electrons, and electron "holes." Any complete study of charge transport must establish the relative concentrations and mobilities of each of these species with respect to all pertinent parameters. The first logical step is to separate the total ionic and electronic contributions.

Ionic transference numbers have been determined by measuring the open circuit potential of a galvanic cell involving a metal oxide between reversible electrodes having different oxygen activities [3,4] and by a coulometric titration of the oxide between reversible electrodes having the same activity [3]. The use of a metal-metal oxide reversible electrode presents difficulties at very high temperatures. One difficulty is in choosing an electrode which is inert with respect to the specimen; any reaction between the electrode and the oxide under study invalidates the result. A second difficulty arises from the fact that the equilibrium between the metal oxide and the gaseous ambient becomes quite rapid, and equilibrium with the electrodes no longer completely defines the oxygen activity profile across the sample. The technique using porous platinum electrodes with controlled oxygen pressures in the gas phase above each face of the sample [4] circumvents most of these difficulties, but requires hermetic seals to a dense, crack-free specimen. Another difficulty, common to all techniques involving a galvanic cell measurement, is that none of these will yield true electronic transference numbers unless measurements are made over a wide range of oxygen pressures, so that the pressure dependence of the electronic conductivity can be established [5].

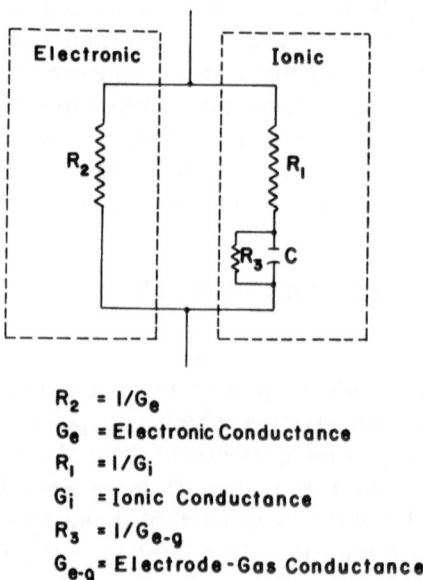

$$R_2 = 1/G_e$$
$$G_e = \text{Electronic Conductance}$$
$$R_i = 1/G_i$$
$$G_i = \text{Ionic Conductance}$$
$$R_3 = 1/G_{e-g}$$
$$G_{e-g} = \text{Electrode-Gas Conductance}$$

Fig. 2. Equivalent circuit of a mixed conductor with electrodes blocking to ions only.

Hebb [6] pointed out that either ionic or electronic transport could be suppressed in a mixed conductor by the proper choice of electrodes, and the contribution of the nonsuppressed species determined. This concept was expanded and extended theoretically by Yakota [7] to include all possible combinations of ionic and electronic electrodes. Electrodes blocking to ions only were used by Danforth and Bodine [8] to study mixed conduction in ThO_2. This technique has been found very useful if a complete analysis of the results is made.

When an electric field below that required for electrolysis or dielectric breakdown is applied across a metal oxide between electrodes which can neither supply or assimilate oxygen ions or metal ions, two mechanisms remain by which an equilibrium current may flow. The first mechanism is electronic conductivity in the metal oxide and is the desired effect, but the second mechanism results from an electrode–gas interaction and can cause errors in the transference number determination.

Since the two electrodes are at different oxygen potentials because of the applied voltage, the oxidation-reduction reactions

$$O^{=}_{\text{surface}} \rightarrow \tfrac{1}{2}(O_2)_g + 2\Theta_{\text{electrode}}$$

$$\tfrac{1}{2}(O_2)_g + 2\Theta_{\text{electrode}} \rightarrow O^{=}_{\text{surface}}$$

(2.1)

between the metal oxide and the gas phase will proceed as thermally-activated processes. This will cause a deviation from blocking electrode conditions, and the current due to this electrode-gas reaction will increase with increasing voltage as long as reaction (2.1) can keep up with the field-induced current of oxygen ions within the metal oxide.

A material exhibiting both ionic and electronic conductivity can be represented by the equivalent circuit shown in Fig. 2, provided its conductivity is measured between electronic electrodes which can either supply or assimilate the ionic species. The ionic resistance R_1 may change with time as the polarization builds up, but this does not affect the determination of transference numbers, since measurements are required only at zero and infinite time, and there is no measurable current through R_1 at infinite time provided the electrode-gas resistance R_3 is large compared to R_2. For the present analysis, it is assumed that the electronic resistance R_2 is not changed by the polarization; the validity of this assumption has been discussed by Danforth and Bodine [8].

If the source current I is constant, then the voltage E across the specimen will have the values

$$E(0) = \frac{R_1 R_2}{R_1 + R_2} I \tag{2.2}$$

$$E(\infty) = \frac{R_2 (R_1 + R_3)}{R_1 + R_2 + R_3} I \tag{2.3}$$

at zero and infinite time. The ratio of initial to final voltages yields

$$\frac{E(0)}{E(\infty)} = \frac{R_1 (R_1 + R_2 + R_3)}{(R_1 + R_2)(R_1 + R_3)} = t_e \left(1 + \frac{R_2}{R_1 + R_3} \right) \tag{2.4}$$

where t_e is the fraction of the total current carried by electrons. If $(R_1 + R_3) \gg R_2$, equation (2.4) reduces to

$$\frac{E(0)}{E(\infty)} = t_e = 1 - t_i \tag{2.5}$$

where t_e and t_i are the transference numbers for electrons and ions respectively. The assumption that the sum of the ionic and electrode–gas resistance is large compared to the electronic resistance can be tested experimentally by considering the conductances at zero and infinite time,

$$G(0) = \frac{I}{E(0)} = \frac{1}{R_2} + \frac{1}{R_1} = G_e + G_i \tag{2.6}$$

$$G(\infty) = \frac{I}{E(\infty)} = \frac{1}{R_2} + \frac{1}{R_1 + R_3} = G_e + \frac{G_i G_{e-g}}{G_i + G_{e-g}} \tag{2.7}$$

where G_e is the electronic conductance, G_i the ionic conductance, and G_{e-g} the electrode-gas conductance. All three conductances in equations (2.6) and (2.7) result from activated processes with, in general, different activation energies. Therefore, a plot of $\log G(\infty)$ versus reciprocal temperature over a sufficiently wide range will be a straight line if the first term in equation (2.7) predominates, but will show a kink if the reverse is true. The possible results can be summarized as follows:

$$G(0) = G_e, \ G(\infty) = G_e \qquad G_e \gg G_i \tag{2.8}$$

$$G(0) = G_i, \ G(\infty) = G_e \qquad G_i \gg G_e \gg G_{e-g} \tag{2.9}$$

$$G(0) = G_i, \ G(\infty) = G_{e-g} \qquad G_i \gg G_{e-g} \gg G_e \tag{2.10}$$

$$G(0) = G_i, \ G(\infty) = G_i \qquad G_{e-g} \gg G_i \gg G_e \tag{2.11}$$

Examples of all four cases will be shown in the section on calcia-stabilized zirconia.

Ionic diffusivities in mixed conductors can be determined if equation (2.5) has been shown to apply. If it is assumed that the ionic conductivity is due to one type of ion only, the diffusion coefficient of that ion can be calculated from the polarization data by means of the Nernst–Einstein relationship. The equation for the diffusion coefficient is

$$D = \frac{lkTl \ [E(\infty) - E(0)]}{Aq^2 \ N \ E(0) \ E(\infty)} \tag{2.12}$$

where l is the sample thickness, A is the sample cross-sectional area, k is Boltzmann's constant, T is the absolute temperature, q is the charge on the diffusing ion, and N is the density of diffusing ions. Equation (2.12) will yield ionic diffusivities even if the majority of the total current is electronic.

The blocking electrode technique has both advantages and disadvantages when compared with other methods for measuring transference numbers in metal oxides. The major advantage for a study of defect structure in metal oxides is that measurements of transference numbers, total conductivity, and thermoelectric power can be made consecutively on the same sample under identical conditions. The major disadvantage is that a careful analysis must be made to ensure that all necessary conditions are satisfied. In addition to the electrode-gas problem discussed above, an error in the $E(0)$ value may arise from dielectric relaxation processes, such as those observed by Sutter and Nowick [9] in NaCl. This effect can be analyzed, if present, by the difference in time constants involved; the relaxation processes require ion movements the order of one atomic spacing, while the blocking electrode polarization takes place over many atomic layers.

Experimentally, the data required for the calculation of transference numbers may be obtained from either time-dependent DC measurements or from a combination of AC and steady-state DC measurements. If the latter method is used, the AC conductivity must be measured as a function of frequency in order to separate ionic and electronic conductivity from relaxation processes. In both cases, an $I(\infty)$ versus $E(\infty)$ curve must be obtained to ensure that the applied voltage is well below that required for electrolysis or dielectric breakdown.

Electrical Conductivity

A determination of the electrical conductivity of a metal oxide as a function of both temperature and oxygen partial pressure pro-

TABLE I

Pressure Dependence of Electrical Conductivity of a
Metal Oxide for Various Possible Defects

$$\sigma \alpha (p_{O_2}^x)_{T=\text{const.}}$$

Oxidation-Reduction Predominates		Thermal Disorder Predominates	
Neutrality Condition	x	Predominant Defect	x
$[V_O^{\cdot}] = n$	$-1/4$	$[V_O^{\cdot}]$ independent of p_{O_2}	$-1/2$
$2[V_O^{\cdot\cdot}] = n$	$-1/6$	$[V_O^{\cdot\cdot}]$ independent of p_{O_2}	$-1/4$
$m[M_i^{m\cdot}] = n$	$-\dfrac{z}{4(m+1)}$	$[M_i^{m\cdot}]$ independent of p_{O_2}	$-\dfrac{z}{4m}$
$[O_i'] = p$	$1/4$	$[O_i']$ independent of p_{O_2}	$1/2$
$2[O_i''] = p$	$1/6$	$[O_i'']$ independent of p_{O_2}	$1/4$
$m[V_M^{m'}] = p$	$\dfrac{z}{4(m+1)}$	$[V_M^{m'}]$ independent of p_{O_2}	$\dfrac{z}{4m}$

z = cation valence and $m = 1, 2, \ldots z$.

vides a convenient and rapid method of investigating the defect
structure.   However, certain precautions must be taken in both
the collection and interpretation of the data before any valid conclusions can be obtained.

The electrical conductivity is the product of charge, concentration, and mobility.   At a constant temperature the mobility
should be relatively insensitive to changes in concentration of
charge carriers, so that changes in concentration should be directly
reflected as changes in conductivity.

The general theoretical approach outlined in Section I is always
applicable and will predict the oxygen pressure dependences of the
electron and hole concentrations for all possible combinations of
predominant defects. The major problem with this general approach
is the mathematical complexity involved; if Schottky disorder,
Frenkel disorder, and oxidation–reduction are allowed for an oxide

MO_2, there are 49 different simplified electroneutrality conditions, each to be used in turn in a simultaneous solution of 18 equations to obtain the analytical solutions for the defect concentrations in all of the 49 potentially possible oxygen pressure ranges. As indicated earlier, each of these 49 analytical solutions generally will not predict a different oxygen pressure dependence for any given defect concentration; in fact, identical dependences would be expected in several regions. This procedure can be simplified at minimum risk by considering only cases near stoichiometry (where thermal disorder predominates) and far removed from stoichiometry (where the neutrality condition involves either electrons or holes). Table I summarizes the pressure dependences for all possible permutations of these two cases. It is not used totally without risk, however, because some of the pressure dependences appearing in Table I might also be obtained in the pressure ranges between near-stoichiometry and extreme oxidation or reduction where this simplification does not apply, and, in addition, even in the ranges where the simplification does apply, an unambiguous dependence is the exception rather than the rule.

Experimentally, the metal oxide must be provided with suitable electrical contacts, and the conductivity must be measured at a known temperature and oxygen partial pressure. The choice of DC or AC measurements is largely determined by the considerations outlined in the section on transference numbers. If AC measurements are used, the conductivity must be measured as a function of frequency to demonstrate that a relaxation process is not influencing the data. The frequency chosen for the measurements should be high with respect to the polarization contribution and low with respect to any relaxation process. With either technique, it is necessary to measure the impedance *in situ* and to eliminate the effects of leakage and surface impedances. This is most conveniently accomplished by introducing a guard ring on the sample (Fig. 3a), which results in a three-terminal network (Fig. 3b) consisting of the sample impedance Z_u, and the leakage and surface impedances Z_1 and Z_2. The lead wires and sample holder must be arranged so that all leakage paths are from Hi to Guard or Lo to Guard. This three-terminal network can then be applied, for example, to a transformer ratio arm bridge (Fig. 3c), a Schering bridge with a guard circuit (Fig. 3d), or a constant current system utilizing a unity gain amplifier as an impedance transformer (Fig. 3e). In all three cases, the impedance between the Hi and Lo terminals is

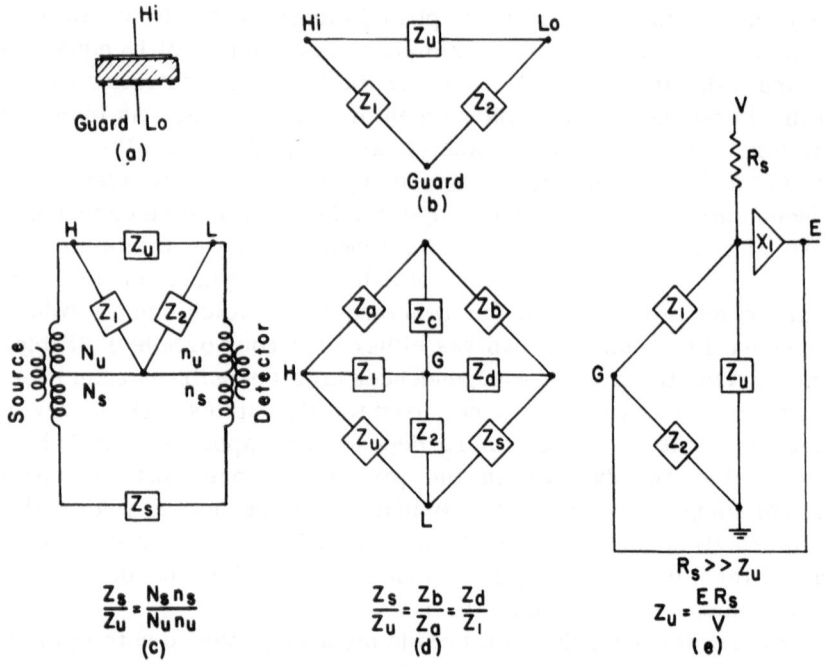

Fig. 3. Three-electrode specimen for conductivity measurements: (a) electrode geometry; (b) equivalent three-terminal network; (c) application to a transformer ratio arm bridge; (d) application to a Schering bridge with a guard circuit; (e) application to a constant current circuit.

measured independent of Z_1 and Z_2. The transformer ratio arm bridge is strictly an AC instrument, but the other two techniques may be used for either AC or DC measurements.

A sample holder configuration which has proven satisfactory is shown in Fig. 4. The only materials of construction in the hot zone are recrystallized alumina and platinum–rhodium alloys. The thermocouples (Pt + 6%Rh vs. Pt + 30%Rh) welded to each electrode are used to measure temperature, temperature difference, and potential difference across the sample in order to calculate thermo-electric power. The two legs of each thermocouple are paralleled for conductivity measurements. It has been found that satisfactory sample electrodes can be obtained by using platinum paste (Engel-hard No. 6082); the samples must be subsequently heated to 800°C in air to remove the organic binder present in the paste.

The oxygen partial pressure in the sample area must be care-fully controlled during conductivity studies. A range of pressures satisfactory for most experiments can be achieved through the use

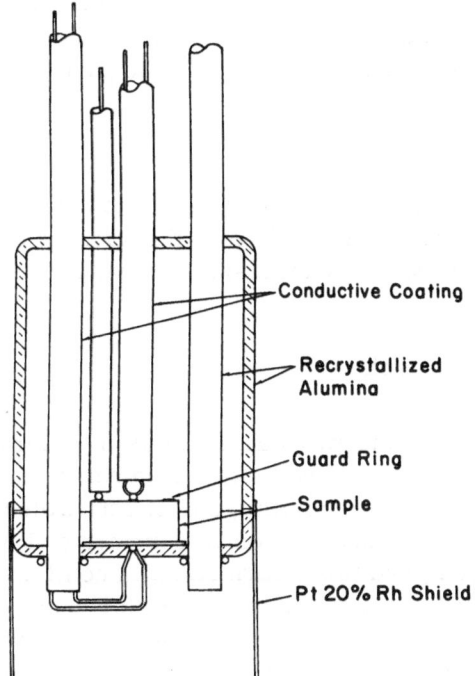

Fig. 4. Three terminal sample holder.

of three techniques. For pressures from $1-10^{-6}$ atm, the total pressure can be reduced and pure oxygen introduced through a variable leak. The system can be operated dynamically in this manner, or statically by backfilling with pure helium to any desired total pressure. Controlled oxygen partial pressures in the intermediate region can be attained by introducing known mixtures of CO_2 and CO and making use of the well-known free energy for the $CO-CO_2$ reaction to calculate the oxygen partial pressure. The same technique using hydrogen and water vapor mixtures can be used to reach the lowest oxygen pressures. With a capability of blending 100:1 ratios, oxygen partial pressures from $1-10^{-24}$ atm can be covered at 1000°C by using these three techniques.

Implicit in the above discussion is the requirement that the metal oxide be in thermodynamic equilibrium with the oxygen partial pressure in the gas phase. To establish the validity of this assumption, the rate of diffusion of oxygen through the lattice must be considered. The choice between single crystal and polycrystalline samples then becomes more difficult. Grain boundaries may adversely influence the conductivity results, but they may also present rapid diffusion paths and so speed equilibrium. The final decision must be based on the characteristics of the particular system under consideration.

Thermoelectric Power

A thermoelectric power (Q) equation of any conductor is obtained by calculating the Thompson coefficient (σ_T) from the appropriate electrical and thermal current density expressions, and then integrating the expression

$$\sigma_T = T \frac{dQ}{dT} \qquad (2.13)$$

For elemental semiconductors, this procedure yields a general expression for Q in terms of Fermi level, forbidden bandwidth, temperature, ratio of electron-to-hole mobility, and effective electron and hole masses [10]. The concentration of carriers can be calculated from Q in any region, and this information, in combination with the conductivity, will yield the mobility. Unfortunately, most metal oxides do not seem to follow the simple band theory of semiconductors; hence, the thermoelectric power equations do not apply. The theory of small polarons can be invoked to arrive at the electrical current density for transition metal oxides, but there is considerable doubt concerning the exact form for the heat of transport resulting from localized charge carriers [11]. Due to the present theoretical uncertainties, the major contribution of a thermoelectric power measurement is to indicate the sign of the majority carrier. The hot junction will be electrically positive for a predominantly n-type material and negative for a p-type, or "hole" conductor.

In principle, the measurement of thermoelectric power is quite simple and straightforward. It is necessary only to measure the Seeback voltage and the temperature gradient across a sample. However, in practice, the techniques for accurately measuring potential difference and temperature difference at high temperatures are quite difficult. The major problems to be overcome are drift of the thermocouple calibrations due to contamination or segragation of impurities, and stray voltages such as those due to emission currents.

Thermobalance

Any change in the concentration of defects due to deviation from stoichiometry will be accompanied by a net change in sample weight. Thus, the most direct way of following a change in defect concentration is gravimetrically. If the weight change is measured as a function of oxygen partial pressure, the predominant types of defects as well as the pressure corresponding to stoichiometry in the metal oxide can be determined.

The weight of any metal oxide sample will always be the sum of the weights of the metal atoms and the oxygen atoms, if the presence of impurities is ignored. If α is the total number of metal atoms in the crystal and is assumed to be constant, the total number of oxygen atoms can be written as

$$\frac{z}{2}\alpha - \varphi V + \frac{z}{2}\theta V - \psi V + \epsilon V$$

where

$$\varphi = [V_O] + [V_O^{\cdot}] + [V_O^{\cdot\cdot}]$$

$$\epsilon = [O_i] + [O_i'] + [O_i'']$$

$$\theta = [V_M] + [V_M'] + [V_M''] + \ldots + [V_M^{z\,'}]$$

$$\psi = [M_i] + [M_i^{\cdot}] + [M_i^{\cdot\cdot}] + \ldots + [M_i^{z\,\cdot}]$$

$$z = \text{valence of the metal}$$

$$V = \text{sample volume}$$

Then the total sample weight W is given by

$$W = \frac{M_M \alpha}{N_0} + \left(\frac{z}{2}\alpha - \varphi V + \frac{z}{2}\theta V - \psi V + \epsilon V\right)\frac{16}{N_0} \tag{2.14}$$

where M_M is the atomic weight of metal and N_0 is Avogadro's number.

The effect of antistructural disorder was neglected in the above treatment. For a metal oxide in which the extent of thermal disorder is small, one of the defect types should predominate under a given set of conditions, and the other three can be neglected. The case where the concentration of vacancies produced thermally is large compared to that due to oxidation–reduction can still be treated by redefining φ, ϵ, θ, and ψ as concentrations in excess of stoichiometry. All of the defect concentrations will have an oxygen pressure dependence which varies with the degree of ionization as discussed previously. For the general nonstoichiometric case at constant temperature, equation (2.14) can be written as

$$W = \frac{a}{N_0} (M_M + 8z) + C p_{O_2}{}^x \qquad (2.15)$$

where x has the possible values given in Table I. Differentiating equation (2.15) with respect to oxygen pressure gives

$$\frac{dW}{dp_{O_2}} = x C p_{O_2}{}^{x-1} \qquad (2.16)$$

Equation (2.16) applies only over oxygen pressure ranges where x is constant. The rate of change of sample weight with respect to oxygen pressure is always positive; for oxygen vacancies or metal interstitials both c and x are negative, while for metal vacancies and oxygen interstitials they are both positive. If the logarithm of both sides of equation (2.16) is taken, the resulting equation is

$$\log \frac{dW}{dp_{O_2}} = (x-1) \log p_{O_2} + \text{constant} \qquad (2.17)$$

Therefore, a plot of $\log dW/dp_{O_2}$ vs. $\log p_{O_2}$ will have a slope of $(x-1)$. The quantity dW/dp_{O_2} is the slope at a point on a plot of W vs. p_{O_2}, but these cannot be plotted directly because of the order of magnitude changes in p_{O_2}. If, however, we let

$$\Delta W = W - W_i \qquad (2.18)$$

where W_i is the initial weight at some fixed p_{O_2}, and observe that

$$\frac{dW}{dp_{O_2}} = \frac{d\Delta W}{dp_{O_2}} = \frac{\Delta W}{p_{O_2}} \frac{d(\log \Delta W)}{d(\log p_{O_2})} \qquad (2.19)$$

then the desired derivative can be obtained from a plot of $\log \Delta W$ vs. $\log p_{O_2}$.

Changes in the predominant defect as a function of oxygen pressure will be reflected by changes in x, and the point at which x changes sign is the true stoichiometric point for the temperature of the measurements. Once the stoichiometric pressure has been established, the gravimetric data can be converted to absolute vacancy concentrations and the precise chemical composition established at any temperature–pressure condition.

A microbalance which has been found to be satisfactory for a study of defect concentration is the Cahn RG Electrobalance. This vacuum microbalance is continuously balanced by a torque motor at

the torsion ribbon; the current to the torque motor is a direct measure of the sample weight, and is monitored on a strip-chart
recorder. The ultimate sensitivity of the balance is 10^{-7} g with a 1-g
sample. The ability to electrically tare out large weight changes
and still maintain the high sensitivity makes this balance ideal for
in situ measurements as a function of temperature and oxygen partial
pressure. The balance in its gas-tight container was mounted on
the atmosphere furnace, and oxygen partial pressure control
achieved by the methods previously described.

III. RESULTS ON SELECTED OXIDES

Calcia-Stabilized Zirconia

The material $Zr_{0.85}Ca_{0.15}O_{1.85}$ is a cubic solid solution with the
fluorite structure. The concentration of oxygen vacancies is chemically controlled by the calcium concentration; one oxygen vacancy
is produced for each calcium atom in the structure. Therefore,
7.5% of the oxygen sites are normally vacant. This oxide has a high
electrical conductivity at high temperatures, which has been
attributed to oxygen ion transport on the basis of measurements of
conductivity as a function of oxygen pressure [12,13] and direct tracer
measurements of oxygen ion diffusion [13].

The $Zr_{0.85}Ca_{0.15}O_{1.85}$ sample labeled ZCS-2 was prepared by dry
mixing weighed amounts of 99.93% ZrO_2 (Wah Chang Corp.), and
99.9% $CaCO_3$ (Eimer and Amend). The mixture was fired at 1000°C
in air for 48 hr, ground in a boron carbide mortar, and cold pressed
at 300 psi. The final sintering of the 0.75-in.-diameter disc sample
was made in air at 1900°C for 6 hr using a gas-fired furnace. The
sample labeled ZCS-6 was made by grinding the ZCS-2 sample in a
boron carbide mortar, adding small quantities of Cr_2O_3, V_2O_5, and

TABLE II

Spectrographic Analysis of Sample ZCS-6

Impurity	Weight Percent
V	0.8
Cr	0.01
Fe	0.1
Al	0.3

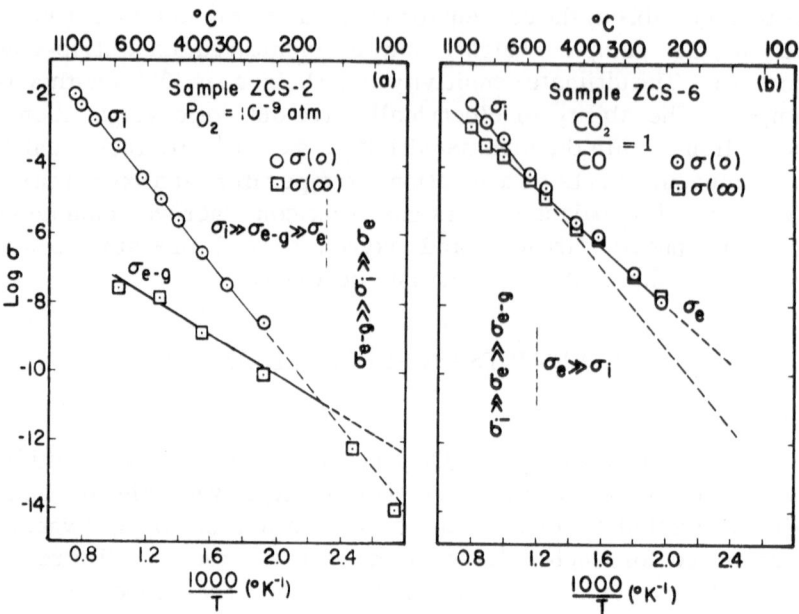

Fig. 5. Electrical conductivity of pure and doped $Zr_{0.85}Ca_{0.15}O_{1.85}$ as a function of temperature.

Fe_2O_3, cold pressing, and firing at 1750°C in air for 6 hr. Results of a spectrographic analysis of ZCS-6 are given in Table II. Since aluminum is not a transition metal, its effect on the electrical properties should be small relative to that of the chromium, vanadium, and iron.

Figure 5a is a plot of both the zero and infinite time conductivities of sample ZCS-2 as a function of temperature at an oxygen partial pressure of 10^{-9} atm. The high-temperature zero-time points lie on a straight line, and yield an activation energy of 1.132 ± 0.015 eV; the variation indicated is the probable error computed from the scatter of points about the least-squares line. This activation energy is in satisfactory agreement with the value found for oxygen ion diffusion [13] (1.22 eV) indicating pure oxygen ion conductivity. The infinite time conductivity values for sample ZCS-2 are represented by two straight lines, the lower temperature line coinciding with the ionic line. This is the behavior predicted by equations (2.10) and (2.11), and indicates that the high-temperature $\sigma(\infty)$ values represent the electrode–gas conductivity. Therefore, the transference number for oxygen ions is 1 for pure $Zr_{0.85}Ca_{0.15}O_{1.85}$ down to at least 100°C.

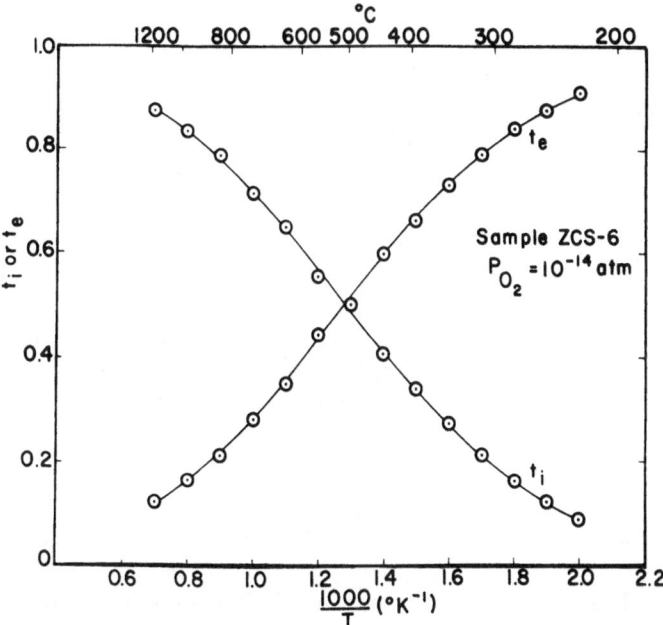

Fig. 6. Transference numbers for sample ZCS-6 as a function of temperature.

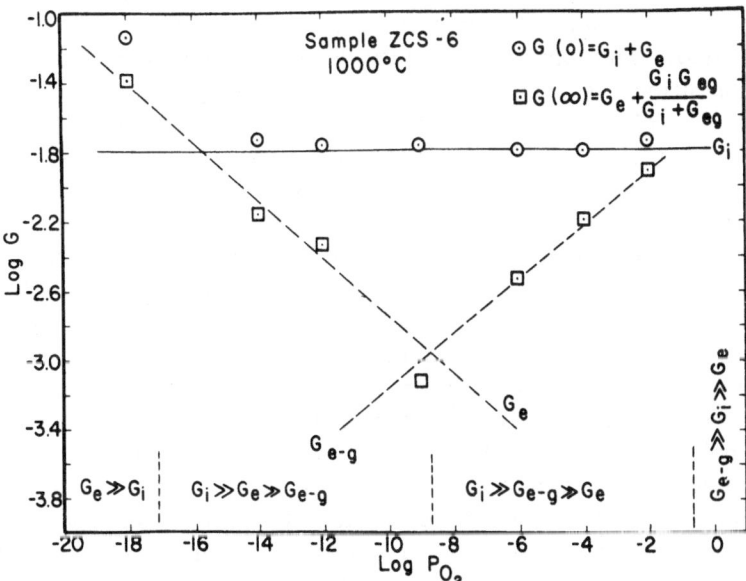

Fig. 7. Electrical conductance of sample ZCS-6 as a function of oxygen pressure at 1000°C.

Figure 5b shows the zero and infinite time conductivities of sample ZCS-6 as a function of temperature in a $CO-CO_2$ mixture which gave an oxygen partial pressure of 10^{-14} atm at 1000°C. The high temperature $\sigma(0)$ points lie on the same straight line as those for sample ZCS-2 (Fig. 5a), and are interpreted as representing predominately ionic conductivity. The $\sigma(\infty)$ data differ considerably from those of sample ZCS-2; the magnitude is greater, and they do not break and follow the ionic line at lower temperatures. This is the behavior predicted by equations (2.8) and (2.9), and indicates that impurity-controlled electronic conductivity has been introduced by the aliovalent transition metal cations added to the lattice. Therefore, equation (2.5) applies to sample ZCS-6 over the entire temperature range, and transference numbers for elections (t_e) and ions (t_i) can be computed. The results of these calculations are shown in Fig. 6.

An oxygen pressure of 10^{-14} atm at 1000°C was chosen to demonstrate the mixed conductivity in sample ZCS-6 as a function of temperature after taking the pressure-dependent data at 1000°C shown in Fig. 7. This plot contains all four conditions predicted by equations (2.8)—(2.11). At very low oxygen pressures, the total sample conductivity is predominantly electronic, but this component decreases with increasing oxygen pressure as is expected for any mechanism in which the electron concentration is defect-controlled. The ionic conductivity, which depends on total vacancy concentration, is constant. As the oxygen pressure is further raised, the electrode-gas conductivity increases as indicated by the increase in $\sigma(\infty)$. This cannot be due to a p-type electronic conductivity because $\sigma(0)$ fails to show a rise at the high pressure end. Based on this analysis, transference numbers for sample ZCS-6 were calculated from equation (2.4) as a function of oxygen pressure at 1000°C. Figure 8 shows the results of these calculations.

Ionic diffusivities were calculated from equation (2.12) for both samples (taking $E(\infty) = \infty$ for sample ZCS-2). The results of these calculations were in good agreement with the direct tracer measurements of Kingery et al. [13] at high temperatures, and follow the extrapolation of their data to temperatures too low for direct diffusivity measurements.

Based on all of the above results, it can be concluded that the cubic stabilized zirconia is a material in which both the defect structure and the electrical conductivity are completely dominated by additions of aliovalent cations. The high oxygen vacancy con-

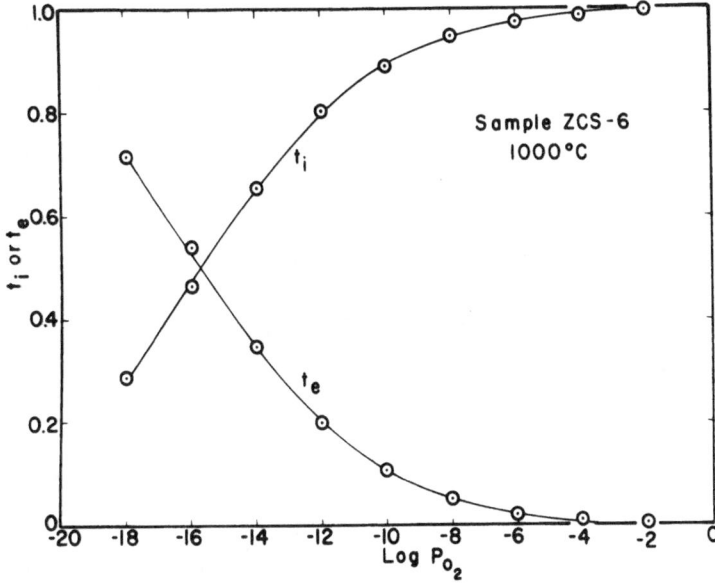

Fig. 8. Transference numbers for sample ZCS-6 as a function of oxygen pressure at 1000°C.

centration and oxygen ion transport are directly related to the calcium content, while appreciable electronic conductivity requires additional transition metal cations to generate noncompensated oxygen vacancies at low oxygen pressures.

Pure Zirconia

A large amount of conflicting information has been published concerning the physical properties of pure ZrO_2. Discrepancies in reports of the crystallographic modifications were pointed out in the recent note by Weber [14] The low-temperature monoclinic and high-temperature tetragonal phases seem well established, but a second very high temperature transformation to a cubic or different tetragonal structure is still in doubt. Rudolph [15] found that monoclinic zirconia was a p-type semiconductor, and that the electrical conductivity varied as the $\frac{1}{5}$ power of the oxygen partial pressure from $1-10^{-6}$ atm. Kofstad and Ruyicka [16], however, found a complex oxygen pressure dependence in both the monoclinic and tetragonal regions. The present work was undertaken to establish the behavior of the electrical conductivity of the monoclinic phase under carefully defined conditions, and to combine these data with thermo-

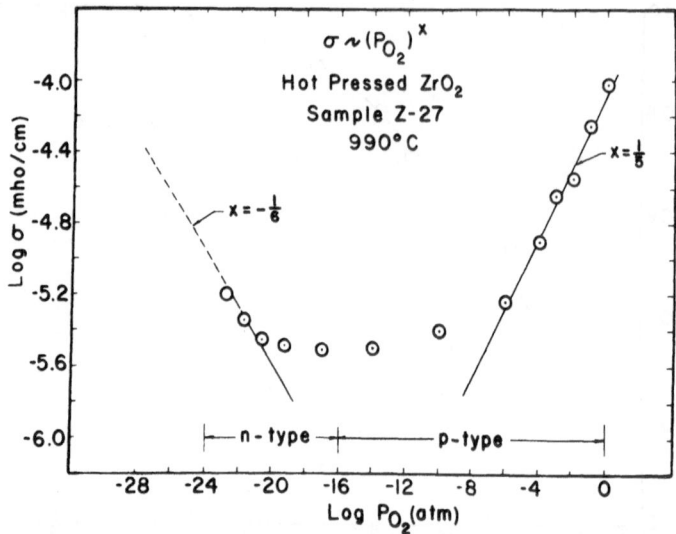

Fig. 9. Oxygen pressure dependence of the conductivity of monoclinic zirconia.

gravimetric measurements to establish the defect structure and to determine the nature of the conduction mechanism.

Dense, polycrystalline samples were prepared by vacuum hot pressing in a tungsten die at 2000°C and 10,000 psi. The starting material contained less than 0.1 wt. % foreign metals. This zirconia was also used in powdered form (320 mesh) for the thermogravimetric analysis.

The conductivity of monoclinic zirconia at 990°C is shown in Fig. 9 as a function of oxygen partial pressure. The data shown are equilibrium points; one hour was found to be sufficient time for equilibrium based on the observation that the conductivity value reached after one hour did not change after twenty-four additional hours at the same conditions. The transition from n-type to p-type conductivity was established by the change in sign of the thermoelectric power. The slope on the high pressure side was computed to be 0.201 ± 0.008 by the method of least squares. The deviation quoted is the probable error calculated from the scatter of points about the least-squares line. The data at the low-pressure end of Fig. 9 are insufficient to define the indicated $-\frac{1}{6}$ slope, but a minimum is definitely present. The broadness of this minimum requires that a third, oxygen-independent term must be added to the analytical representation of the data in Fig. 9 regardless of the exact dependence at the low-pressure end. Assuming the $-\frac{1}{6}$ dependence in the n-type region, the total conductivity at 1000°C can be represented by

$$\sigma_{1000°} = 8.5 \cdot 10^{-5} \, p_{O_2}^{1/5} + 1.1 \cdot 10^{-9} \, p_{O_2}^{-1/6} + 3.2 \cdot 10^{-6} \qquad (3.1)$$

The first term, which is predominant in the high oxygen pressure region, can be attributed to the formation of completely ionized zirconium vacancies. The precision of the experimental data does not permit the possibility of either a $\frac{1}{4}$ or $\frac{1}{6}$ dependence, and fortunately the $\frac{1}{5}$ dependence can arise from only one mechanism, namely, the complete ionization of zirconium vacancies where the concentration of vacancies produced by Schottky disorder is small compared to that resulting from oxidation. The second term in equation (3.1) which is predominant at very low oxygen pressures, may be due to completely ionized oxygen vacancies, but the experimental data are insufficient to unequivocally define the slope. The third, oxygen pressure-independent term is even less well understood. This term may be due to ionic conductivity, but preliminary polarization studies of transport numbers do not substantiate an ionic contribution of this magnitude. Impurity conduction by either a hopping process or promotion to a normally empty $5s$ or $4d$ band is also a possibility.

The data shown in Fig. 9 suggest that stoichiometry for zirconia at 1000°C occurs at an oxygen pressure near 10^{-16} atm, and that the metal-to-oxygen ratio is less than 0.5 at all higher oxygen pressures. However, conclusions concerning the metal-to-oxygen ratio based on electrical data alone must be qualified because of three assumptions which are basic to the interpretation. The first is that the pressure corresponding to the minimum can be obtained from the intersection of the linear portions of the curve, and the second is that the electron and hole mobilities are equal, so that the minimum corresponds to the $n = p$ point. The fact that the thermoelectric power, which depends on carrier concentrations but not mobility, changes sign at the pressure at which the minimum conductivity occurs gives credence to the assumption of equal mobilities. The third assumption is that $\delta = 0$ at the $n = p$ point, and this was shown to be generally not true for most oxides. Since, as shown in Fig. 1, the $\delta = 0$ point generally occurs at lower pressures than the $n = p$ point in MO_2 oxides, the stoichiometric point in ZrO_2 should occur at an oxygen pressure somewhat less than 10^{-16} atm.

The weight change of a zirconia sample was measured as a function of oxygen pressure at 1000°C. The analysis outlined in Section II was applied to the data and the results are shown in Fig. 10. The points can be represented by a straight line having a slope of $-4/5$, which is the slope required for $x = 1/5$. Therefore, the thermogravimetric results with zirconia at 1000°C can also be interpreted in terms of fully-ionized zirconium vacancies being the predominant defect at high oxygen pressures.

The weight changes ΔW should be a linear function of $p_{O_2}^{1/5}$. This prediction is followed within the scatter of the data as shown in Fig. 11. At the particular reference weight chosen ($\Delta W = 0$ at $p_{O_2} = 0$), ΔW is given by

$$\Delta W = C p_{O_2}^{1/5}$$
$$\Delta W = 1.73 \cdot 10^{-4} p_{O_2} \tag{3.2}$$

where ΔW is measured in grams and p_{O_2} in atmospheres.

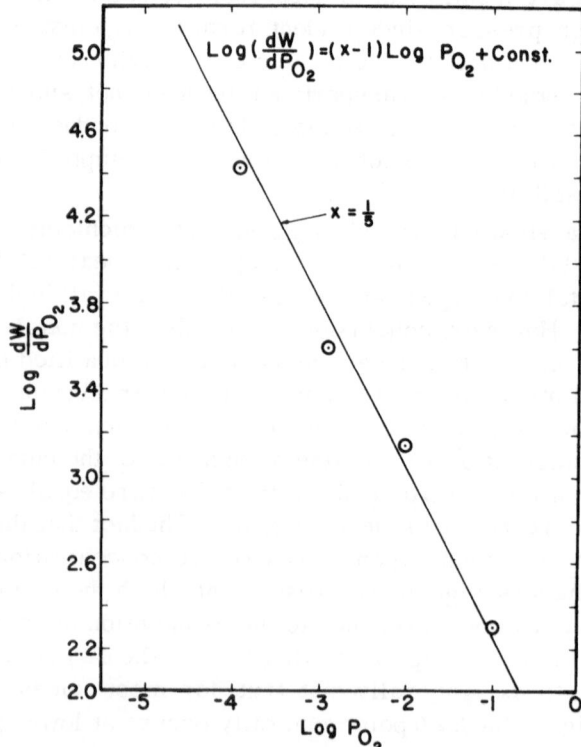

Fig. 10. Plot of equation (2.17) for ZrO_2 at 1000°C.

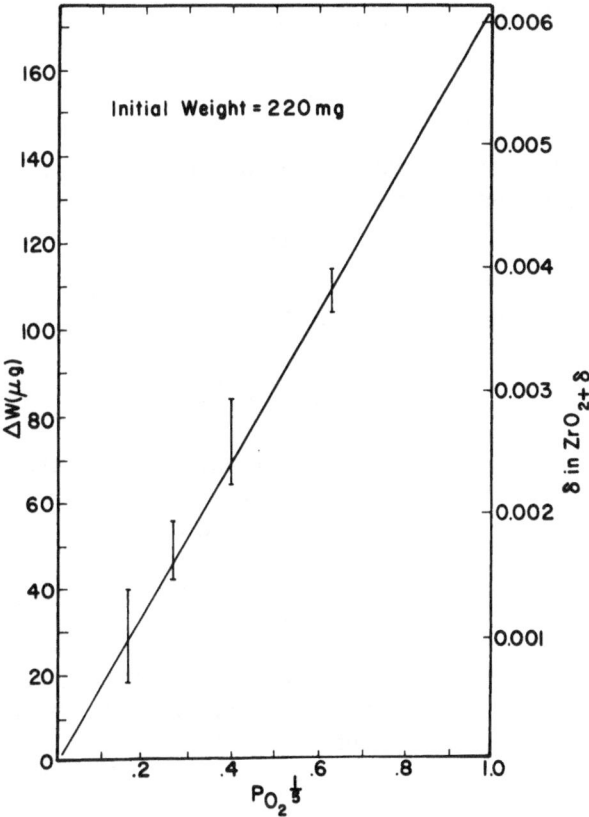

Fig. 11. Weight change and deviation from stoichiometry for ZrO_2 at 1000°C.

A determination of the degree of oxygen rich nonstoichiometry in zirconia is possible by combining the electrical and thermogravimetric results. From Fig. 9 it was concluded that, at 1000°C, stoichiometry occurs near an oxygen pressure of 10^{-16} atm. At this point the concentration of oxygen vacancies is equal to twice the concentration of zirconium vacancies, and equation (2.14) reduces to

$$W_{stoich.} = \frac{123.22 \, \alpha}{N_0} \tag{3.3}$$

If this is taken as the initial weight W_i, then at high oxygen pressures equations (2.14), (2.18), and (3.3) can be combined to give

$$\Delta W = \frac{32 V \theta}{N_0} = \frac{32 W \theta}{\rho N_0} \tag{3.4}$$

where ρ is the density. The concentration of zirconium vacancies is then

$$\theta = \frac{\rho N_0 \Delta W}{32W} \qquad (3.5)$$

If the formula for zirconia is written as $ZrO_{2+\delta}$ and zirconium vacancies are considered the predominant defect, δ will be given by

$$\delta = \frac{2\theta M}{\rho N_0} \qquad (3.6)$$

where M is the molecular weight of zirconia. The combination of equations (3.5) and (3.6) gives

$$\delta = \frac{M \Delta W}{16W} = \frac{123.22 C p_{O_2}^{1/5}}{16W} \qquad (3.7)$$

For convenience ΔW was referenced to zero pressure rather than 10^{-16} atm; the error introduced is negligible at high oxygen pressures $(p_{O_2} > 10^{-8})$. Insertion of the sample weight and the value of C determined from Fig. 11 gives

$$\delta = 6 \cdot 10^{-3} p_{O_2}^{1/5} \qquad (3.8)$$

where p_{O_2} is measured in atmospheres. Equation (3.8) should be valid at oxygen pressures large relative to 10^{-16} atm. Fig. 11 is a plot of δ as well as ΔW as a function of $p_{O_2}^{1/5}$.

The addition of the thermogravimetric results to the electrical results can also shed additional light on the mechanism of charge transport. At 1000°C and high oxygen pressures, zirconia is a p-type semiconductor so the conductivity will be given by

$$\sigma = p|e|\mu \qquad (3.9)$$

where e is the electronic charge and μ the mobility. The mobility must be independent oxygen pressure, at least at high pressures where the conductivity and weight change show the same $\frac{1}{5}$ dependence on p_{O_2}. If μ varied with oxygen pressure, the plot of log σ vs. log p_{O_2} (Fig. 9) would not be linear. Since $p = 4\theta$, equation (3.9) becomes

$$\sigma = 4\theta|e|\mu \qquad (3.10)$$

The substitution of the value of σ from equation (3.5) gives

$$\sigma = \frac{\rho N_0 \Delta W|e|\mu}{8W} \qquad (3.11)$$

The assumptions are best satisfied if only changes in σ and ΔW within the high pressure region are considered. Equation (3.11) may be rewritten in terms of changes and solved for the mobility:

$$\mu = \frac{8}{N_0|e|} \frac{W}{\rho} \frac{(\sigma_1 - \sigma_2)}{(\Delta W_1 - \Delta W_2)} \quad (3.12)$$

For monoclinic zirconia at 1000°C, a change in oxygen pressures from 10^{-4} to 10^{-1} atm gave a mobility calculated from equation (3.12) of

$$\mu_{1000°} = 1.4 \cdot 10^{-6} \text{ cm}^2/V\text{-sec} \quad (3.13)$$

This mobility is much too small for hole conduction in any simple band; the band width would have to approach zero. The "small" polaron model [17] of activated hopping motion offers the best possibility for accommodating a mobility of this magnitude, but neither the theoretical nor experimental results are sufficient to permit a quantitative evaluation. It would be desirable to determine the temperature dependence of the mobility, but the tetragonal phase transformation occurs at higher temperatures, while at lower temperatures the time required for equilibration with the atmosphere becomes prohibitively long.

IV. NONTRANSITION METAL OXIDES

While the preceding section described two specific cases in which the characterization of the defect structure in transition metal oxides could be deduced experimentally, and while many similar studies of other transition metal oxides have been published, surprisingly little is known about the defects present in wide band gap nontransition metal oxides as a function of impurity content and deviation from stoichiometry or the mechanisms by which electronic charges move in these materials. Several problems, however, may immediately be anticipated from the results that have been obtained so far. For example, as suggested previously, it may be expected that the presence of transition metal impurities will play a predominant role in determining the concentrations of defects and their oxygen pressure dependences in the nearly stoichiometric oxides. The electron and hole conductivities in MgO were found by Mitoff [18] to be intimately associated with the iron content in the specimens, and similar results have been found recently in Al_2O_3 by Harrop and Creamer [19]. Elusive relaxation effects, apparently associated with trapped electronic charges, but strongly influenced by thermal history and the oxygen pressure in which the specimens have been quenched, have been observed in MgO by Southgate [20] using internal friction techniques and in Al_2O_3 by the authors using dielectric measurements. While it is entirely reasonable that the electrons and holes generated by reduction or oxidation in the nearly stoichiometric oxides should be associated with the transition metal impurities, a fact which we denote by saying that the impurities change valence, it is certainly not clear how these impurities control the observed conductivity, since a theory for low-mobility electron conduction in the dilute impurity concentration case is not available. It would seem then that at this point any characterization of the defects present based solely on electronic transport measurements must be quite speculative.

Even the measurement of ionic transport in these nontransition metal oxides containing small concentrations of transition metal impurities may be quite confusing in the oxygen pressure range where the oxide is nearly stoichiometric. Davies [21], for example, has reported the observation of both magnesium and oxygen vacancy transport in MgO, but found the oxygen vacancy transport at higher oxygen pressures than those at which he found the magnesium vacancy transport. He also noted that other observers have found similar phenomena. While such inversions of the intuitively anticipated behavior might be accounted for by arguments similar to those presented in the latter part of Section I, it is again clear that the characterization of the defect concentrations present in these materials by transport measurements alone may be quite difficult.

It has already been demonstrated by Wertz, Orton, and Auzins [22] that the transition metal impurities and the vacancies in their various degrees of ionization in MgO can be identified and that changes in the degrees of ionization with oxygen pressure can be studied by electron-spin resonance measurements. In fact, as they have noted, a nontransition metal oxide, such as MgO, is ideal for such a study because of the nonparamagnetic nature of the predominant magnesium and oxygen isotopes. The combination of this approach to the direct characterization of the defects present in this and similar oxides with transport property and optical measurements should lead to a rapidly increasing understanding of these materials.

REFERENCES

1. F. A. Kroger and H. J. Vink, "Relations Between the Concentrations of Imperfections in Crystalline Solids," in: Solid State Physics, Vol. III, edited by F. Seitz and D. Turnbull, 307-435, Academic Press, New York, 1956.
2. G. Brouwer, Philips Research Repts. 9:366 (1954).
3. C. Wagner, Proc. Intern. Comm. Electrochem. Thermodynam. and Kinet.; 7th Meeting, 1955 (Butterworth Publications Ltd., London, 1957).
4. S. P. Mitoff, J. Chem. Phys. 36:1383 (1962).
5. H. Schmalzried, Z. Elektrochem. 66:572 (1962).
6. M. H. Hebb, J. Chem. Phys. 20:185 (1952).
7. I. Yokota, J. Phys. Soc. Japan 16:2213 (1961).
8. W. E. Danforth and J. H. Bodine, J. Franklin Inst. 260:467 (1955).
9. P. H. Sutter and A. S. Nowick, J. Appl. Phys. 34:734 (1963).
10. V. A. Johnson and K. Lark-Horovitz, Phys. Rev. 92:226 (1953).
11. M. M. Chadda and A. P. B. Sinha, Indian J. Pure Appl. Phys. 1:161 (1963).
12. K. Kiukkola and C. Wagner, J. Electrochem. Soc. 104:379 (1957).
13. W. D. Kingery, J. Pappis, M. E. Doty, and D. C. Hill, J. Am. Ceram. Soc. 42:393 (1959).
14. B. C. Weber, J. Am. Ceram. Soc. 45:614 (1962).
15. J. Rudolph, Z. Naturforsch. 14a:927 (1959).
16. P. Kofstad and D. J. Ruzicka, J. Electrochem. Soc. 110:181 (1963).
17. G. L. Sewell, Phys. Rev. 129:597 (1963).
18. S. P. Mitoff, J. Chem. Phys. 31:1261 (1959).
19. P. J. Harrop and R. H. Creamer, Brit. J. Appl. Phys. 14:335 (1963).
20. P. D. Southgate, Armour Research Foundation Report ARF A:945 (1962).
21. M. O. Davies, J. Chem. Phys. 38:2047 (1963).
22. J. E. Wertz, J. W. Otron, and P. Auzins, J. Appl. Phys. 33:322 (1962).

Direct Observation of Radiation Damage In Molybdenum

J.D. Meakin and A. Lawley

The Franklin Institute Laboratories
Philadelphia, Pennsylvania

The primary objective of the research is to study the effect of neutron irradiation, both at ambient and elevated temperatures, on the structure of molybdenum, and to compare and contrast with previous observations on copper and nickel. The defect structures produced by neutron irradiation at doses up to approx. 10^{20} *nvt*, at temperatures from ambient to 600°C, are being studied using transmission electron microscopy. For low doses approx. 10^{18} *nvt* at 600°C, large prismatic dislocation loops (average diameter 1100 A) were observed on planes approximately parallel to {111}, and having the normal slip vector $1/2 <111>$. Diffraction contrast experiments were carried out on these loops which proved conclusively that these were formed by the collapse of interstitial aggregates. By comparison, ambient temperature irradiation to this dose level produced a structure with a high density of spots which were not readily resolvable as loops. Specimens have also been irradiated to approx. 10^{20} *nvt* at ambient temperature and have been found to contain a higher density of the unresolved damage. The mechanism of the formation of the loops and their interaction with glide dislocations is discussed.

I. INTRODUCTION

Studies of irradiation damage using transmission electron microscopy have led to the conclusion that, in many cases, clustering of point-defects takes place, thereby producing prismatic dislocation loops. In the case of copper, gold, and molybdenum, neutron irradiation at ambient pile temperatures produces small loops less than approx. 100 A in diameter. This size limitation has restricted a detailed study of the crystallography and exact nature (vacancy or interstitial aggregation) of the loops. By comparison, large dislocation loops with diameters in excess of 1000 A have been produced in a number of metals [1,2] by bombardment with fission fragments or heavy ions. Such loops have all been shown to be interstitial in nature; the sign of the dislocation is determined by observing the form of the dislocation contrast as the diffraction conditions are varied [3].

An insight into the nature and crystallography of the irradiation

irradiation has a profound influence on the nature of the damage, and may completely suppress the formation of visible structures. Whereas copper irradiated at 70°C to a dose of approx. 10^{17} nvt fast flux contains many resolvable loops, nickel irradiated to a similar dose at the same temperature shows no resolvable structures. However, by irradiating nickel at 200°C, visible damage is formed which is very similar in appearance and density to that produced in copper at the lower temperature. These temperatures correspond to the same homologous temperature ($T/T_M \simeq 0.26$) for each metal, so that the diffusion distances should be somewhat similar in each case. It therefore appears likely that the diffusion distance or point-defect mobility is a controlling factor in the formation of prismatic loops, which would be directly influenced by the temperature existing at the time of irradiation. The present paper describes a series of irradiation experiments carried out on molybdenum, a metal having a much higher melting point than copper or nickel, and possessing a body-centered cubic structure. In addition to neutron irradiation at ambient temperatures, irradiations were made at 600°C which corresponds to $T/T_M \simeq 0.3$. It was anticipated that at this temperature a comparatively modest dose of approx. 10^{18} nvt would give rise to visible damage, and this in fact proved to be the case.

II. EXPERIMENTAL PROCEDURE

Single-crystal specimens for irradiation were grown from General Electric vacuum arc-cast stock using the electron beam vertical floating zone technique (Lawley [4]). Analysis of the zoned molybdenum gave an overall purity of approx. 99.995% with interstitial impurity levels of carbon approx. 13 ppm by weight, and oxygen, nitrogen, and hydrogen approx. 3 ppm by weight total. Disc specimens of approx. 0.8 mm thickness were cut from the zoned rod using a diamond slitting wheel. In order to maintain the specimens at 600°C during the irradiation, a small resistance heating furnace was constructed which fitted into a standard welded aluminum irradiation can. The specimens were stacked inside the furnace with alternate spacers of silica, and the assembly was then sealed in the can under helium at a pressure of about $\frac{1}{4}$ atm (Fig. 1). Thermocouple and power leads were made into the can using a multipoint glass-metal seal. The irradiation was carried out in a light water moderated reactor with the specimens maintained at temperatures of 600 ±20°C. A nickel dosimeter sample was

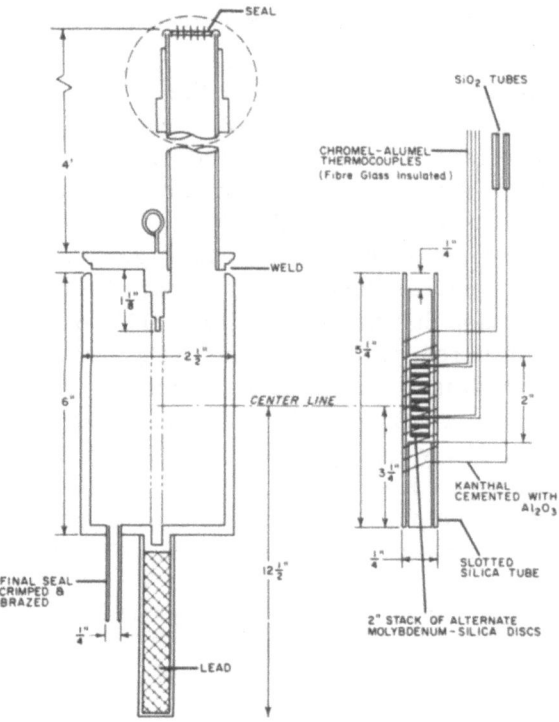

Fig. 1. Aluminum capsule used for high-temperature irradiation.

used to monitor the fast flux which was measured to be $2.8 \cdot 10^{12}$ neutrons/cm^2-sec. This gave a total fast flux dose of $8 \cdot 10^{17}$ nvt for the 87-hr period of irradiation. As a control experiment, several of the unirradiated discs were annealed at 600°C for 90 hr in order to investigate the influence, if any, of the thermal treatment alone.

Specimens suitable for thin-film transmission electron micros-copy were prepared from the discs using the jet polishing tech-nique [5]. A two-axis goniometer stage holder was utilized in the Philips 100B electron microscope, with the latter operating at 100 kV. In this way, contrast effects due to selected operating reflections could be obtained. When necessary, a standard X-ray backreflection Laue photograph was taken of the thinned sample actually mounted in the microscope holder. This enabled the speci-men orientation to be determined in a manner free from the 180° ambiguity inherent in electron diffraction.

III. ELECTRON MICROSCOPE OBSERVATIONS

A representative micrograph of the damage produced by the 600°C irradiation is given in Fig. 2. Loops are visible which are both perpendicular and inclined to the (011) foil plane. The perpendicular loops appear as lines approximately parallel to the [2$\bar{1}$1] and [21$\bar{1}$] directions, indicating that the loops lie on the {111} planes. This conclusion is supported by the appearance of the inclined loops, which in general show a long axis parallel to the [01$\bar{1}$] direction. Confirmation that the loops do lie on {111} planes was obtained when foils prepared parallel to the (111) plane were examined. These foils contained three sets of inclined loops that could be imaged in strong contrast. There were also some approximately circular loops that were apparently parallel to the foil and which could only be imaged very weakly. A weak contrast intensity for loops parallel to the plane of the foil is consistent with the Burgers vector of these loops, as is discussed later. There is therefore no

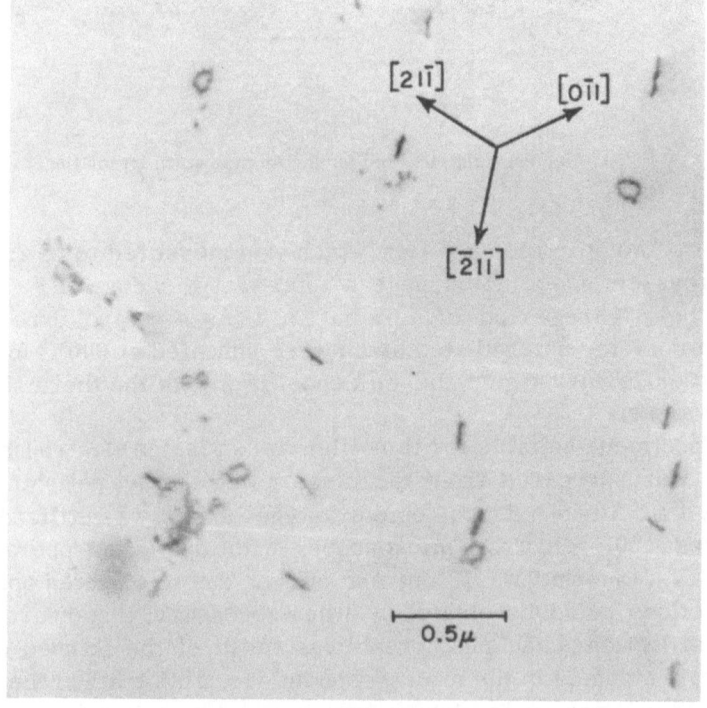

Fig. 2. Prismatic loops with plane of foil (011).

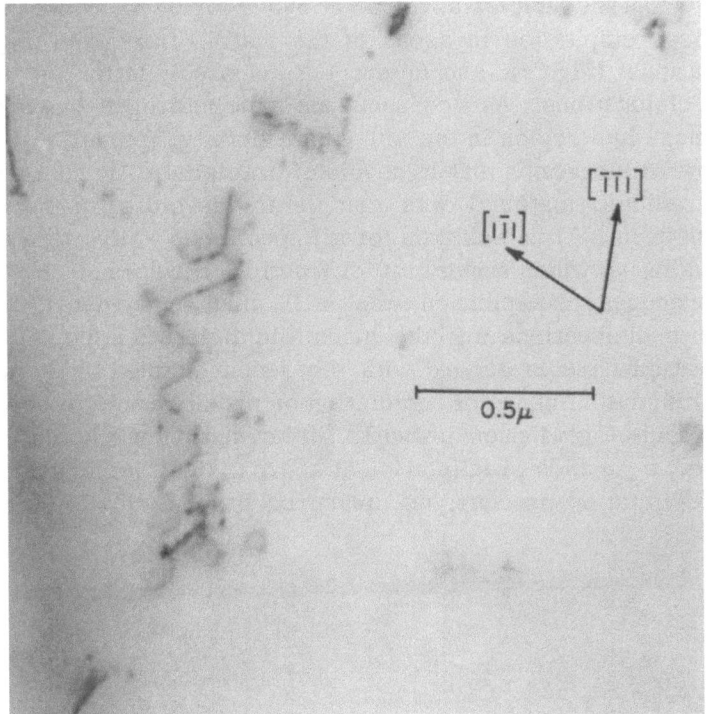

Fig. 3. Prismatic dislocation loops and helices. Plane of foil (011)

doubt that the loops tend to lie on {111} planes, but nevertheless readily detectable deviations between the plane of the loops and the crystallographic {111} planes exist.

The majority of loops have diameters close to the average size of 1100 A; the smallest resolvable loop was about 100 A in diameter and occasionally loops as large as 2000 A were observed. From previous experience with thinned molybdenum specimens (Lawley and Gaigher [6]), the foil thickness may be taken as 2000 A, which gives a loop density $2 \cdot 10^{13}/cm^3$. The defect concentration required to form this loop density is $4.3 \cdot 10^{-5}$.

An interesting feature quite commonly observed was the conversion of in-grown dislocations into helices, presumably by the adsorption of point-defects produced by the irradiation. Figure 3 shows both loops and helices in the same region of an (011) foil; as expected, the helices are parallel to $\langle 111 \rangle$ directions. In the example shown the particular $\langle 111 \rangle$ directions are in the plane of the foil, and therefore considerable lengths of helices are visible.

The control samples annealed at 600°C for 90 hr showed homogeneous precipitation in areas of the matrix free from in-grown dislocations (Fig. 4), and heterogeneous precipitation on the in-grown dislocations. As a consequence of the heterogeneous precipitate, there is a region in the immediate vicinity (approx. $1\,\mu$) of the in-grown dislocations that is devoid of precipitate. By comparison, the irradiated material was completely devoid of homogeneous precipitation. This was true for all specimens which were examined using a wide variety of diffracting conditions. However, heterogeneous precipitation was still much in evidence on both in-grown dislocations and the prismatic dislocation loops. These observations are in accord with the recent studies by Damask et al. [7] on the influence of irradiation on precipitation processes in iron. Thus, irradiation prevents the formation of a homogeneous metastable carbide precipitation at approx. 160°C. The kinetics of the precipitation process, as measured by electrical resistivity,

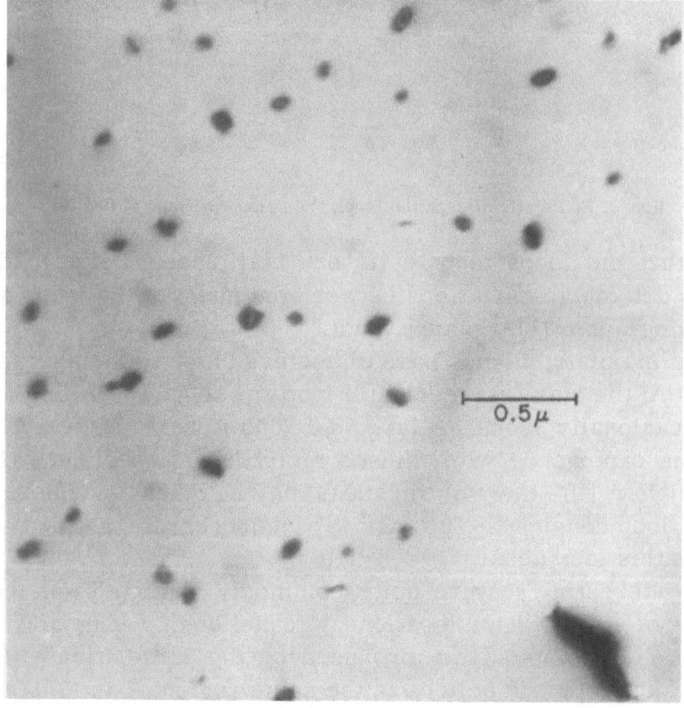

Fig. 4. Homogeneous precipitation in an unirradiated sample annealed at 600°C for 90 hr. Plane of foil (011).

calorimetry, and transmission electron microscopy, indicate that the interstitial impurity carbon atoms are trapped at the radiation-produced point-defects. It is only at higher temperatures that the impurities can leave these traps.

IV. DETERMINATION OF THE BURGERS VECTOR OF THE LOOPS

There are a number of possible planes in a bcc lattice on which point-defects may be expected to aggregate and collapse to give a prismatic loop. Consequently, a number of Burgers vectors appear possible for the observed prismatic loops; for a quantitative analysis, a series of contrast experiments were carried out. The fact that the loops lie on {111} planes is at least suggestive that they have an $(a/2)\langle 111\rangle$ Burgers vector, this of course corresponding to the lowest energy dislocation in a bcc lattice.

Figure 5 shows a group of loops as seen using the various operating reflections available from an (011) foil. Three sets of loops are distinguishable; two sets of loops perpendicular to the foil on planes $(1\bar{1}1)$ and $(11\bar{1})$ and inclined loops on the planes $(\bar{1}11)$ and (111). The latter two sets of loops are indistinguishable in projection, since they are equally inclined to the [011] foil normal. In Table I the values of $g \cdot b$ are listed for the three sets of loops on the assumption that the loops are pure edge and hence have an $(a/2)\langle 111\rangle$ Burgers vector. The vector g is perpendicular to the active plane of reflection. Comparison of Table I and the image intensities observed in Fig. 5 provides convincing evidence that the loops do have Burgers vectors parallel to $\langle 111\rangle$ and hence

$$b = \frac{a}{2} \langle 111\rangle$$

TABLE I

Magnitudes of $g \cdot b$ for the Operating Reflections Available from an (011) Foil

Loop	b	g [200]	g [01$\bar{1}$]	g [21$\bar{1}$]	g [2$\bar{1}$1]
A	$(\frac{1}{2})[1\bar{1}1]$	1	1	0	2
B	$(\frac{1}{2})[11\bar{1}]$	1	1	2	0
C	$(\frac{1}{2})[\bar{1}11]$ or $(\frac{1}{2})[111]$	1	0	1	1

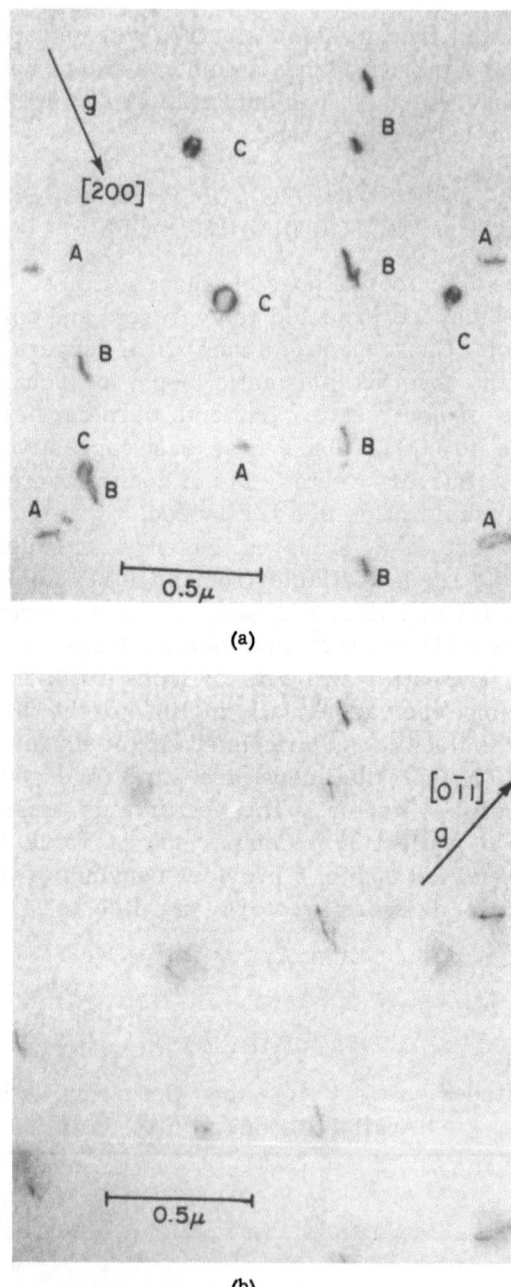

(a)

(b)

Fig. 5. Prismatic dislocation loops in an (011) foil. Loops A are on (1Ī1), loops B on (11Ī), and C on (Ī11) or (111).

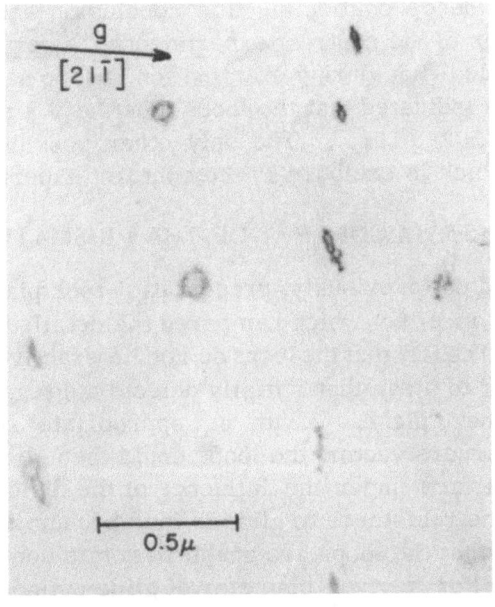

(c)

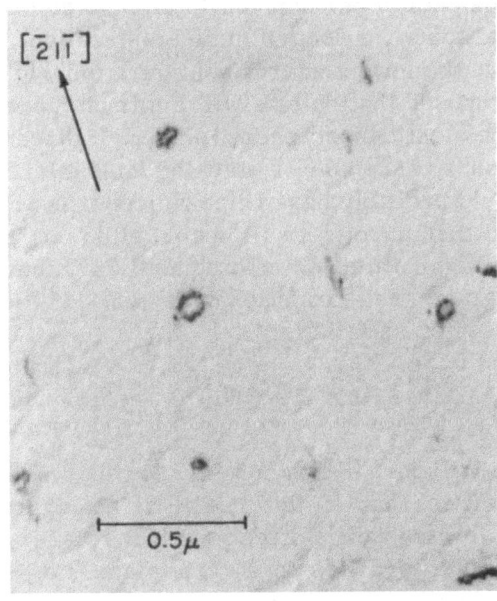

(d)

Fig. 5 continued.

Additional evidence confirming this conclusion was obtained after the completion of the contrast experiments. A sequence of micrographs revealed that during examination, a loop has slipped out of the foil. This indicated that the loops must have a perfect Burgers vector, i.e., $(a/2) \langle 111 \rangle$. The only other possibility would be a $\langle 100 \rangle$ type which is excluded by the contrast experiments.

V. CRYSTALLOGRAPHY OF THE PRISMATIC LOOPS

As pointed out previously, precipitation took place on the prismatic loops, an effect which hampered the detailed analysis of the loop shapes. The fact that the loops do not lie exactly on {111} planes is interpreted to mean that initially defects aggregate and collapse on some other plane.   With an appropriate shear to give an $(a/2)\langle 111 \rangle$ Burgers vector, the loops could then glide along a prismatic slip surface under the influence of the dislocation line tension. Since the resistance to glide in molybdenum is considerable, it is possible that the loops are unable to rotate completely into the {111} plane.   For a given diameter of glide cylinder, the shortest loop length is realized in the completely edge configuration. However, since the energy per unit length of edge dislocation is higher than that of a screw, a loop of mixed dislocation, i.e., inclined to {111}, may give the lowest energy configuration. In the bcc lattice, it has been proposed that defects will aggregate in the {110} planes, these being the most densely packed planes. If direct collapse without offset occurs, a region of stacking fault is created which in a bcc lattice would probably have a high fault energy. For this reason, it is expected that an offset will occur, either during the collapse of the defect aggregate, or subsequently by the nucleation of a partial dislocation. The formation of an $(a/2)\langle 111 \rangle$ loop may therefore be represented as follows:

$$\frac{a}{2} \langle 110 \rangle \; + \; \frac{a}{2} \langle 001 \rangle \; \rightarrow \; \frac{a}{2} \langle 111 \rangle$$

Kuhlmann-Wilsdorf [8] has considered this type of loop production and observes that if the initial defect aggregate has sides parallel to the close-packed directions in the plane of collapse, then each side of the loop lies in a {110} plane, i.e., the normal slip planes. Specifically, the close-packed directions in a {110} plane are two $\langle 111 \rangle$ directions and a $\langle 100 \rangle$ direction. These directions include angles of 125° and 110° and, therefore, an almost perfect hexagon can form. A rotation of 35° is necessary to bring a loop from the {110} plane to the {111} plane normal to the Burgers vector, and, in the case cited, this can be accomplished by slip in {110} slip planes. In the final pure edge configuration on a {111} plane, the sides of the loops are parallel to ⟨112⟩ directions which make included angles of 120°. Any given $(a/2)\langle 111 \rangle$ loop can be created in three different ways as follows:

$$\frac{a}{2}[011] \ + \ \frac{a}{2}[100] \rightarrow \frac{a}{2}[111]$$

$$\frac{a}{2}[101] \ + \ \frac{a}{2}[010] \rightarrow \frac{a}{2}[111]$$

$$\frac{a}{2}[110] \ + \ \frac{a}{2}[001] \rightarrow \frac{a}{2}[111]$$

Rotation of a regular loop from each of the above {110} planes to the (111) plane would give loops that are shortened along each of the three ⟨112⟩ directions in the (111) plane. A close study of Fig. 6b reveals that the loops on either of the inclined planes, for example the primed loops on (111), show more than one configuration, but the resolution of the loop sides is insufficient to allow conclusive analysis. Further speculation on the mechanism of formation of the loops is unjustified at this stage. A possible complication not mentioned previously is that the loops may grow by climb following rotation to the {111} planes. The present understanding of the process of loop formation is such that a consideration of alternative modes of aggregation or collapse would not add materially to the discussion.

VI. THE SPECIFIC NATURE OF THE PRISMATIC LOOPS

The diffraction contrast method of determining whether a prismatic loop corresponds to an aggregation of vacancies or interstitials is now well established [1]. In bright-field viewing, the dark contrast due to a dislocation is displaced from the true position of the dislocation axis. By changing the diffraction conditions, it is possible for a dislocation loop to appear smaller or larger as the contrast occurs on the inside or outside of the true loop position. To establish the nature of a loop, it is necessary to know in which sense the loop is inclined in the foil and the direction of rotation necessary to bring the Bragg fringe associated with the

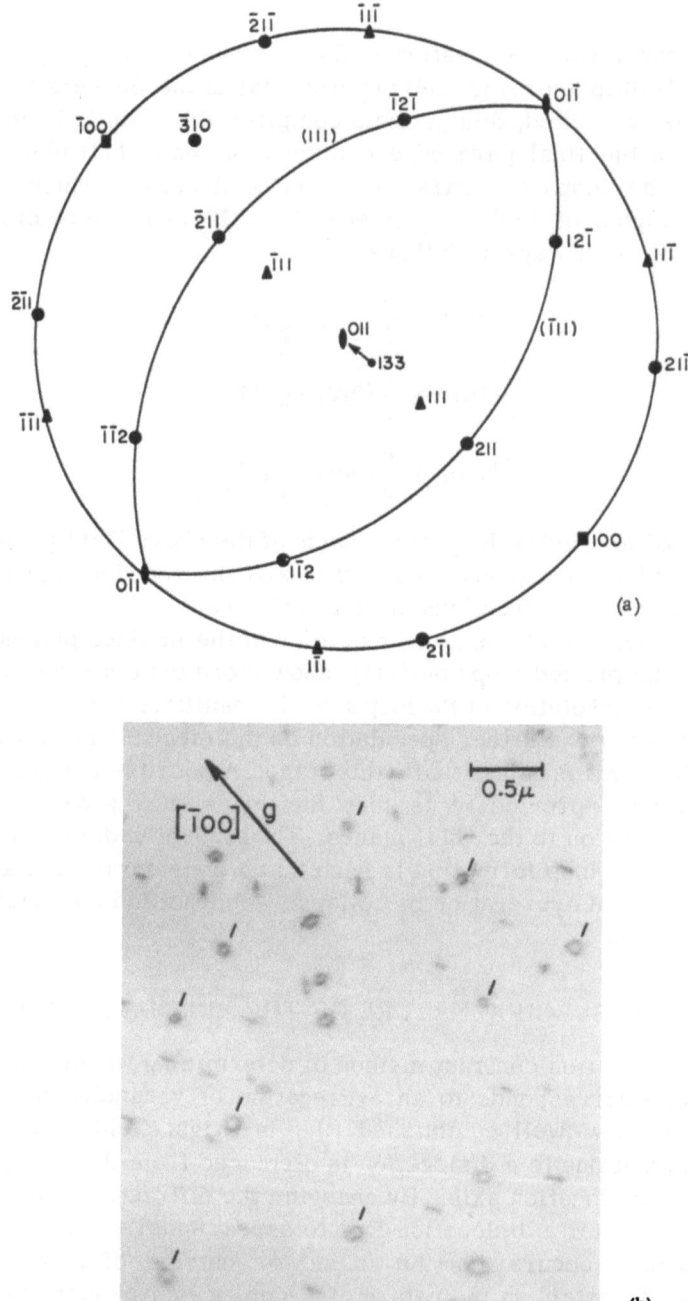

Fig. 6. Stereogram and micrographs on an (011) foil showing method used for determination of the character of the loops.

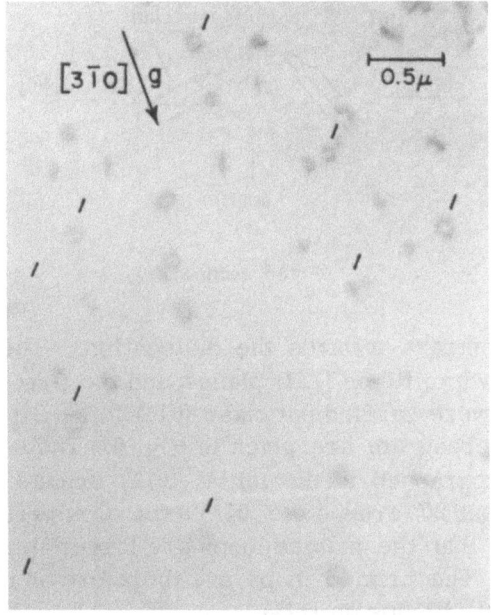

(c)

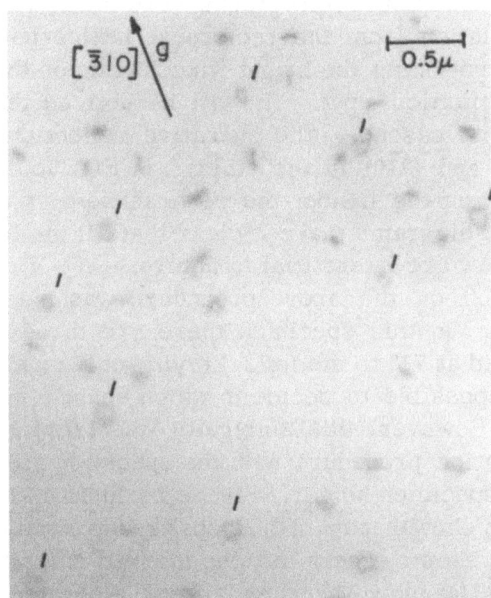

(d)

Fig. 6 continued.

Fig. 6 continued.

dislocation contrast towards the dislocation. The loops studied have been shown to lie on {111} planes, and the first experiments to be described were carried out on an (011) foil. Explanatory drawings and a stereogram are given in Fig. 6. The same area of the foil was photographed in the initial (011) orientation and after a rotation of about 20° around the [01$\bar{1}$] axis. Comparison of Figs. 6b and 6d shows that the primed loops are larger in projection after the rotation. The primed loops are therefore on the (111) planes and the unmarked loops on ($\bar{1}$11).

The nature of the loops can now be determined by comparing Figs. 6c and 6d. Both micrographs were taken with $\bar{s}$ set positive ($\bar{s}$ is the deviation from the reciprocal lattice point to the Ewald sphere) by positioning the bright Kikuchi line on the outside of the operating diffraction spot. It will be noticed that although $\bar{s}$ is positive in both cases, g (the operative reflection) is ($\bar{3}$10) in the one position and (3$\bar{1}$0) in the other. In Fig. 6c the primed loops have contrast on the inside, the reverse being true for the loops on ($\bar{1}$11). The diagrams make it clear that all the inclined loops in the field of view are interstitial in nature.

A variation on the above procedure was made using a (111) oriented foil. In this specimen there are three sets of loops on planes inclined at 71° to the foil. Relying only on electron diffraction, it is impossible to decide in which sense a given {111} plane is inclined. However, this ambiguity was eliminated by using an X-ray orientation procedure with the specimen still in position on the Philip's specimen holder. Figure 7 contains all the necessary information to show that again the loops are interstitial in character. All the loops shown except the one marked A can be concluded to lie on the ($\bar{1}$11) plane as they appear elongated along the [0$\bar{1}$1] direction. Figure 7b confirms this conclusion as the loops here show minimal contrast, the operating reflection [0$\bar{1}$1] being perpendicular to [$\bar{1}$11] and hence $g \cdot b = 0$. Loop A is on the (11$\bar{1}$) plane and clearly shows the reverse contrast to the loops on ($\bar{1}$11) in keeping with opposite inclinations of these two planes.

VII. INTERACTION OF GLIDE DISLOCATIONS WITH PRISMATIC LOOPS

A further example of the effect of the actual temperature of irradiation is afforded by observations on deformed material. Several of the irradiated discs (approx. 10^{18} *nvt* fast flux at 600°C) were deformed (approx. 5% reduction in thickness) by rolling parallel to the (011) plane. Transmission micrographs of these samples showed a predominance of screw dislocations along predicted directions, consistent with this symmetrical slip orientation, as in Fig. 8a. There is no marked effect on the glide dislocation configurations following 600°C irradiation and no significant change in hardness, (VDH 227 ±·4). These observations are consistent with the comparatively low total loop density ($2 \cdot 10^{13}$ cm^{-3}) and large interloop spacing (approx. 0.3μ). The micrographs are very similar to those obtained in unirradiated molybdenum strained in tension, as in Fig. 8b.

A completely different situation exists following neutron irradiation at ambient temperature (40°C) and subsequent tensile deformation (Mastel et al. [9]). In this case, damage is in the form of black spots 75–100 A in diameter with a density $\gtrsim 10^{15}$ loops/cm^3. The glide dislocations have to sweep out or channel through the clusters of point-defects, and this gives rise to a significant increase in yield strength and a region of prolonged easy-glide. Again, in the case of nickel (Wilsdorf [10]), irradiation at ambient temperature does not give rise to discernible damage; nevertheless, there is a considerable increase in strength, with the glide dislocations heavily jogged. Thus, it is the very small, closely-spaced, defect aggregates and individual point-defects that have the major effect on the mobile dislocations.

VIII. DISCUSSION

The observable effect of the neutron irradiation of molybdenum at 600°C is a comparatively low density (approx. 10^{13}/cm^3) of large prismatic dislocation loops. Similar neutron doses of 10^{18} *nvt* in metals, such as copper or gold, at ambient temperatures produce about 10^{15} loops/cm^3 with an average size less than 100 A. It has

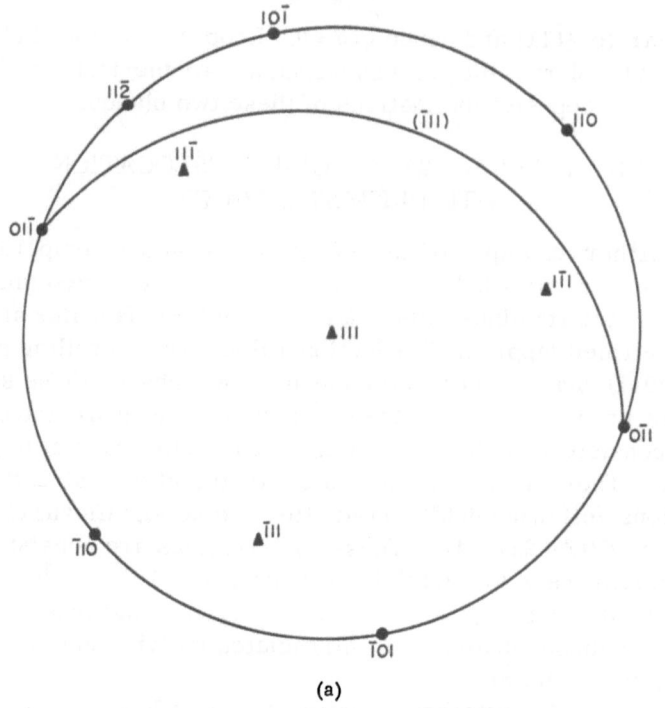

(a)

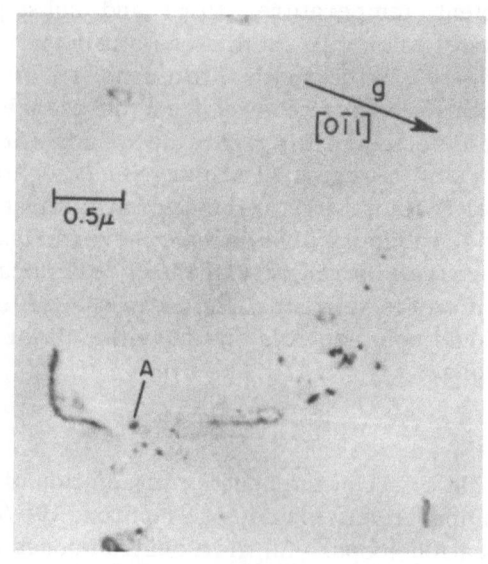

(b)

Fig. 7. Loop analysis using a (111) foil.

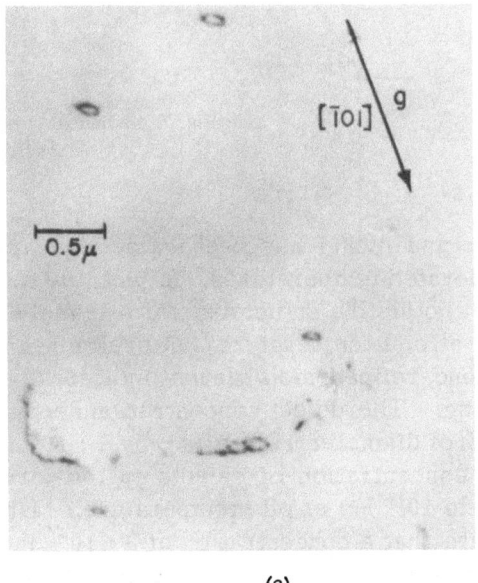

(c)

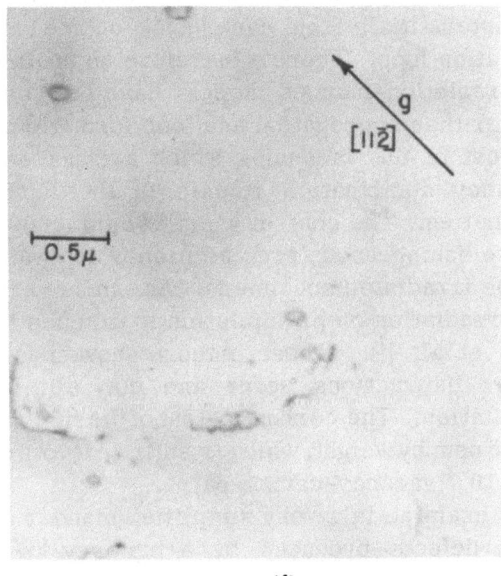

(d)

Fig. 7 continued.

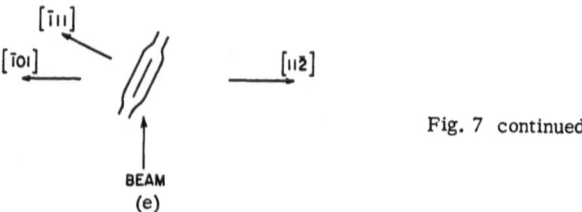

Fig. 7 continued.

damage in copper and nickel has been gained by carrying out the irradiations at elevated temperatures. In fact, the temperature of been reported that no visible structures form in molybdenum after a dose of 10^{18} nvt at pile temperature, but preliminary results obtained with the zone refined molybdenum indicate that visible defects are occurring. The defect concentration necessary to form $2 \cdot 10^{13}/cm^3$ loops of diameter 1100 A is only $4 \cdot 10^{-5}$. This is far smaller than the concentration of defects in the form of loops in copper irradiated to 10^{18} nvt at pile temperature. Makin and Manthorpe [11] estimate that a concentration of $3 \cdot 10^{-4}$ interstitials is in the form of loops in copper. The interstitial concentration is also small in comparison to the number of incident neutrons or the corresponding number of primary knock-ons. The total interstitial atom content in the molybdenum is $2 \cdot 10^{18}/cm^3$, which means that only about four interstitials from each knock-on are retained in the form of a dislocation loop. There is therefore no doubt that a large fraction of the radiation-induced defects have been annihilated by mechanisms other than aggregation and loop formation.

Probably most of the vacancies which are not annihilated by interstitial-vacancy combination remain in the lattice bound to interstitial impurities. The control samples held at 600°C for 90 hr showed extensive homogeneous precipitation which was completely suppressed in the irradiated specimens. An extensive investigation of the effect of irradiation on precipitation in iron has been carried out by Damask et al. [7]. These authors show conclusively that impurity-vacancy interactions occur and may eliminate certain modes of precipitation. The carbon content of the molybdenum used here was about 15 ppm by weight, which is sufficient to form an atom fraction of $1.2 \cdot 10^{-4}$ vacancy-carbon pairs.

We will now examine in a very simplified manner the probable behavior of the defects produced by a primary knock-on. The purpose of the analysis is to show that aggregation of interstitials into loops is likely, but that little or no clustering of vacancies can be expected during normal irradiation. The structure arising from

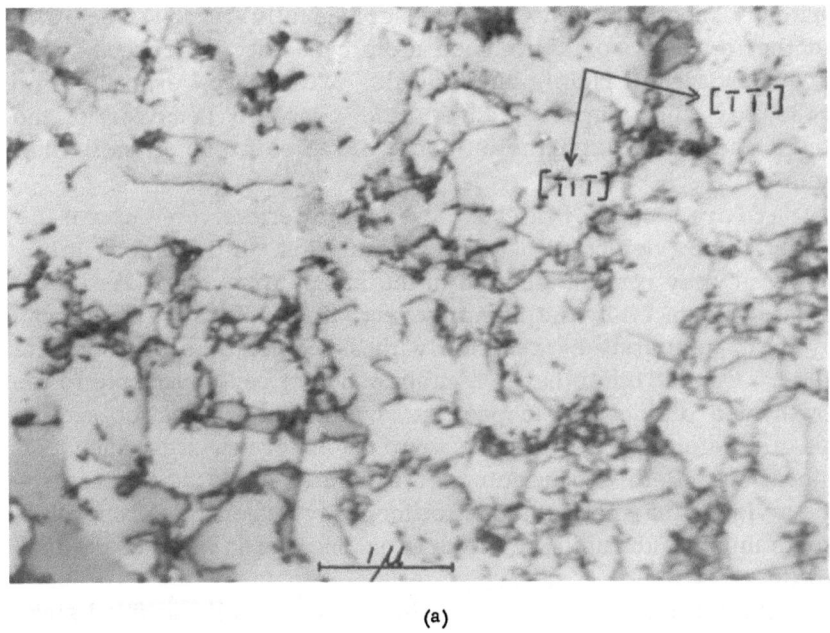

(a)

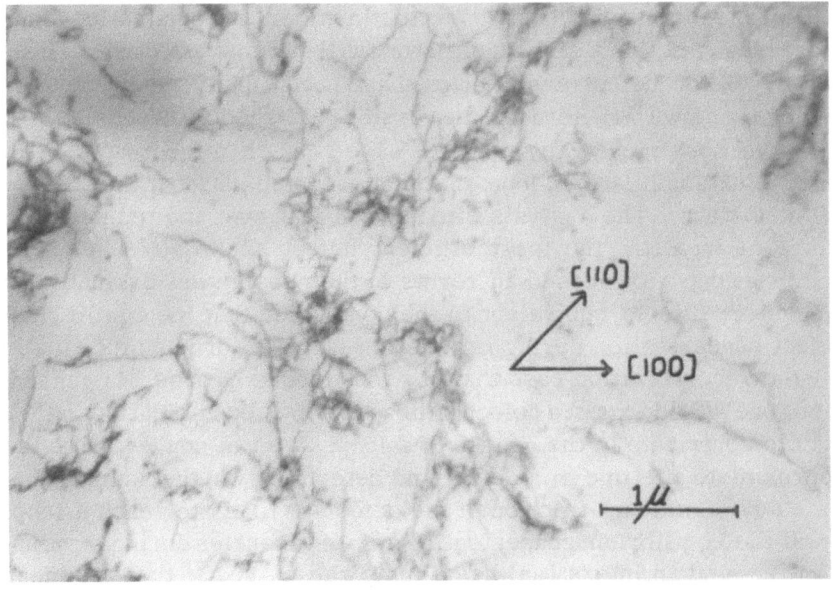

(b)

Fig. 8. (a) Molybdenum irradiated at 600°C, followed by 5% compression. (011) foil.
(b) Unirradiated molybdenum, strained 8.3% in tension. (001) foil.

a primary knock-on cannot yet be described fully, but it is generally held that a region of diminished atomic density is produced surrounded by a shell rich in interstitials. We will now examine, in a necessarily approximate manner, the probable behavior of the defects and particularly the influence of temperature on their final status.

There are a number of events which may effectively control the final state of a defect produced by irradiation. If the defect is immobile, it may remain isolated in the lattice or perhaps remain in a cluster produced by the primary event. If immobile the defect can only be annihilated by reaction with a mobile defect of opposite nature. On the other hand, a mobile defect can contribute to the growth of a dislocation loop or be annihilated by reaction with an opposite defect or be absorbed into an in-grown sink, such as a dislocation. Before continuing, it should be stressed that to produce a visible prismatic dislocation loop, say larger than about 50 A, defects must be diffusing to the original nucleus so as to cause loop growth.

At the initiation of irradiation there are virtually no fixed sinks for defects, so that defect-annihilation must be mainly by interstitial-vacancy collisions. As the number of prismatic loops increases, the fixed-sink annihilation will become effective. These two types of defect-annihilation have been analyzed by several authors, and we will here use the version published by Wechsler [12]. The two possible idealized cases have been examined: namely, by fixed-sink annihilation alone and defect recombination in the absence of fixed sinks. The analysis cited shows that over the temperature range of interest, the most efficient defect annihilation is by recombination. Expressed in terms of the parameters evaluated by Wechsler, this is equivalent to the conclusion that the equilibrium defect concentration during irradiation is orders of magnitude lower for the recombination case than for the fixed-sink case. In general then, one would expect a defect to disappear by recombination rather than to contribute to the growth of a loop. We can now evaluate the approximate lifetime of a defect and determine whether during this time any significant fraction of the defects is likely to reach a loop.

If the equilibrium concentration of free vacancies during irradiation is c and an interstitial has a jump rate ν_i sec^{-1}, then the mean lifetime of a defect will be of the order $\frac{1}{2}(c\nu_i)^{-1}$. The interstitial has the highest mobility, and therefore the lifetime of either a vacancy or an interstitial is controlled by ν_i.

We can now evaluate the mean diffusion distance of each defect before annihilation by recombination is likely. The mean diffusion distance in three dimensions of a defect of diffusion coefficient D after a time t is $\bar{x} = \sqrt{6Dt}$. For our purposes, a sufficiently accurate value of D is $\frac{1}{6}\nu a^2$ where ν is the jump rate. The values of $\bar{x}$ are then as follows:

$$\bar{x} \simeq 0.2\, a c^{-\frac{1}{2}} \qquad \text{for interstitials}$$

and

$$\bar{x} \simeq 0.2\, a (\nu_v/\nu_i)^{\frac{1}{2}} c^{-\frac{1}{2}} \text{ for vacancies}$$

The value of $\bar{x}$ for interstitials could clearly have been arrived at in a more direct manner. If there is a concentration c of vacancies, an interstitial must make about $c\nu_i^{-1}$ jumps before annihilation and the root-mean-square diffusion distance would then be of the order $ac^{-\frac{1}{2}}$.

The magnitudes of $\bar{x}$ for copper and nickel are calculated using the c values published by Wechsler. The neutron flux used for irradiating the copper, nickel, and molybdenum specimens was in the range 10^{11} to 10^{12} nvt, and we therefore use the value of c corresponding to a defect production rate K of 10^{-8} atom fraction/sec. For molybdenum, we use Wechsler's analysis and take the activation energy of motion of an interstitial as 0.4 eV after Nihoul [13]. The calculated values are listed in Table II for the metals and temperatures of interest. It is seen that for interstitials, $\bar{x}$ is of order 10^{-5} to 10^{-4}, whereas, except for molybdenum at high temperatures, the mean diffusion distance of a vacancy is at the most about one interatomic distance. If loops of density n cm^3 are to grow, then the appropriate defects must diffuse over distances of

TABLE II

The Influence of the Temperature of Irradiation on c, the Radiation Induced Vacancy Concentration, and on $\bar{x}$ the Mean Diffusion Distance of a Defect

Temperature, °K	Molybdenum			Copper or Nickel	
	300	900	1250	300	500
c	10^{-7}	$5 \cdot 10^{-10}$	$3 \cdot 10^{-10}$	10^{-9}	$3 \cdot 10^{-10}$
$\bar{x}$ interstitial (cm)	$2 \cdot 10^{-5}$	$2 \cdot 10^{-4}$	$3 \cdot 10^{-4}$	$2 \cdot 10^{-4}$	$3 \cdot 10^{-4}$
$\bar{x}$ vacancy (cm)	$6 \cdot 10^{-13}$	$8 \cdot 10^{-7}$	$5 \cdot 10^{-6}$	$4 \cdot 10^{-11}$	$3 \cdot 10^{-8}$

the order $n^{-1/3}$ and it is quite certain that only interstitials can cause loop growth. In view of the extensive simplifications used, the values of $\bar{x}$ calculated are in satisfactory agreement with the observed range of loop densities, namely, 10^{12} to $10^{15}/cm^3$.

In contrast, vacancies are virtually immobile keeping in mind their probable lifetime, and one may anticipate very little clustering other than that arising directly from the primary displacement event. It would seem just possible that in molybdenum, significant vacancy clustering could occur at high temperatures. However, as the temperature is raised, the degree of vacancy supersaturation diminishes until the vacancy concentration is equal to the thermodynamic equilibrium concentration. This temperature is estimated to be 500°K for copper and 1250°K for molybdenum, and, of course, above these temperatures, vacancy clustering would not occur. At 500°K in copper, $\bar{x}$ for vacancies is only 10^{-8} cm; but, at 1250°K in molybdenum, $\bar{x}$ is 10^{-5} cm and therefore, as stated above, vacancy clustering may be just possible in molybdenum.

ACKNOWLEDGMENTS

This research was supported by the United States Atomic Energy Commission. The authors are grateful to W. Hepfer for skillful technical assistance and to Dr. I. G. Greenfield of the University of Delaware for many stimulating discussions. The authors are indebted to American Machine and Foundry for permission to use their irradiation facilities at the Industrial Reactor Laboratory, Plainsboro, New Jersey.

REFERENCES

1. D. J. Mazey, R. S. Barnes, and A. Howie, Phil. Mag. 7:1861 (1962).
2. B. C. Masters, Nature 200:254 (1963).
3. P. B. Hirsch, A. Howie, and M. J. Whelan, Phil. Trans. Roy. Soc. (London) A257: (1960).
4. A. Lawley, Electronics 32:39 (1959).
5. P. R. Strutt, Rev. Sci. Instr. 32:411 (1961).
6. A. Lawley and H. L. Gaigher, Phil. Mag. 8:1713 (1964).
7. H. Wagenblast, F. E. Fujita, and A. C. Damask, Acta Met. (in press).
8. D. Kuhlmann-Wilsdorf, Phil. Mag. 3:125 (1958).
9. B. Mastel, H. E. Kissinger, J. J. Laidler, and T. K. Bierlein, J. Appl. Phys. 34:3637 (1963).
10. H. G. F. Wilsdorf, Phys. Rev. 3:172 (1959).
11. M. J. Makin and S. A. Manthorpe, Phil. Mag. 8:1725 (1963).
12. M. S. Wechsler, ASTM Bull., S. Tech. Pub. No. 341, 86 (1963).
13. J. Nihoul, Phys. Stat. Solidi 2:308 (1962).

Dislocations in Deformed Beryllium

V. V. Damiano and M. Herman

The Franklin Institute Research Laboratories
Philadelphia, Pennsylvania

High-purity beryllium single crystals were deformed in tension for basal and for prism slip. Some of the work-hardening mechanisms operating at various stages were deduced from observations of dislocations in foils cut from the bulk crystals and from slip lines observed on the surface. Observation of long-edge pairs and edge dipoles in foils cut from crystals deformed in stage I for basal slip suggests that screws have high mobility on the basal plane in stage I and that the crystals exhibit very little hardening. In stage II the presence of numerous edge boundaries was associated with the onset of a rapid rate of work hardening just prior to failure. Three stages of hardening were observed for crystals deformed for prism slip. In stage I the observation of a predominance of screw dislocations suggested that screw dislocation intersections with the grown-in networks had to occur, producing jogs in the screws which acted to impede the motion of the screw dislocations. In stage II complex interactions produced complicated tangled masses of dislocations. In stage III the onset of duplex slip produces stable low-angle boundaries as a result of dislocation interactions along intersections of glide planes.

I. INTRODUCTION

In recent years attention has been drawn to the study of the plastic properties of hexagonal metals. It has been observed that at room temperature the conditions for twinning, the number of twinning systems, and the choice and number of slip systems differ among them. The explanation of what determines the choice of the operating slip systems is lacking.

In particular, the prediction as to which slip plane should operate based upon the c/a ratio of the lattice has met with little success. Magnesium, zinc, and cadmium with c/a ratios 1.624, 1.856, and 1.886, respectively, have as their primary slip mode $\{0001\}$ $\langle 11\bar{2}0 \rangle$. Hafnium, zirconium, and titanium with c/a ratios less than that for ideal close packing (1.633) have as the primary slip mode $\{10\bar{1}0\}$ $\langle 11\bar{2}0 \rangle$. This change in the primary slip mode seems reasonable since as the c/a ratio decreases, the basal plane is less closely packed. Unfortunately, beryllium with the smallest c/a ratio of 1.568 has $\{0001\}$ $\langle 11\bar{2}0 \rangle$ as the primary slip mode;

thus, it is apparent that considerations of axial ratio alone are insufficient in determining which is the primary slip system of hexagonal crystals.

The primary slip system in these metals only refers to that system which will operate at the lowest value of shear stress. Other slip systems can operate at higher stresses. The ratio of the shear stress of these secondary systems to the primary one can vary from 2 to 1 to over 1000 to 1. Explanations for this stress variance have only been made for some of the hexagonal metals in temperature regions where creep can occur.

Gilman [1] reported that the ratio of the activation energy for prism glide for the zinc and cadmium was in the ratio of the elastic shear constants on the prism planes.

Study of the thermally activated mechanism of prismatic slip in magnesium single crystals led Flynn, Mote, and Dorn [2] to conclude that the activation of cross slip in magnesium on the prism plane was based on the thermally activated cross slipping of dissociated screw dislocations in the basal plane. This implies that a work-hardening mechanism peculiar to the prism plane can account for the greater stress required to obtain prism slip with respect to basal slip at high temperatures.

In beryllium (as in titanium [3]) both prism and basal flow occur at room temperature. No satisfactory explanation for the difference in plastic flow properties between the two operating slip systems has been presented.

The purpose of the present study was to compare the shear–stress vs. shear–strain behavior of beryllium single crystals deformed for basal slip with those deformed for prism slip and to compare the dislocation configurations and slip lines observed at various levels of strain. An attempt was made to derive from the observations an explanation for the observed differences in work-hardening characteristics associated with the different modes of deformation.

For convenience, the dislocation interactions and slip in beryllium are presented by a system of notation similar to that proposed by Thompson [4] for face-centered cubic metals. The case of hexagonal metals is more complex, but use can be made of the double tetrahedron [5], as shown in Fig. 1, to describe some of the more important slip systems encountered in hexagonal metals. The double tetrahedron constitutes $\frac{1}{6}$ of the close-packed hexagonal cell shown and can be used to describe the slip planes and Burgers

(a) $ABB'A'=(01\bar{1}0)$, $BCC'B'=(1\bar{1}00)$, $ACC'A'=(\bar{1}010)$

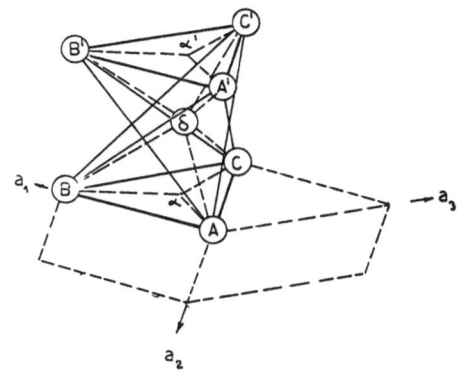

(b) $AB'C'=(\bar{1}101)$, $BC'A'=(10\bar{1}1)$, $CB'A'=(0\bar{1}11)$
$A'BC=(1\bar{1}01)$, $B'CA=(\bar{1}011)$, $C'BA=(01\bar{1}1)$

(c) $A\delta B'=(\bar{2}112)$, $B\delta C'=(1\bar{2}2)$, $C\delta A'=(1\bar{2}12)$
$A'\delta B=(2\bar{1}\bar{1}2)$, $B'\delta C=(\bar{1}\bar{1}22)$, $C'\delta A=(\bar{1}2\bar{1}2)$

Fig. 1. Double tetrahedron notation showing (0001), {10$\bar{1}$0}, {10$\bar{1}$1}, and {11$\bar{2}$2} slip planes.

vectors associated with the dislocations participating in the deformation process. The important slip planes described are the basal plane designated by the plane ABC; the three primary prism planes designated by $ABB'A'$, $BCC'B'$, and $ACC'A'$; the three secondary prism planes designated $A\alpha\alpha'A'$, $B\alpha\alpha'B'$, and $C\alpha\alpha'C'$; six primary pyramidal planes designated by the planes $AB'C'$, $BC'A'$, $CB'A'$, $A'BC$, $B'CA$, and $C'BA$; and six secondary pyramidal planes designated by $A\delta B'$, $B\delta C'$, $C\delta A'$, $A'\delta B$, $B'\delta C$, and $C'\delta A$.

Slip occurs in beryllium at room temperature on the ABC plane and the vectors AB, BC, and CA describe the $\frac{1}{2}\langle 11\bar{2}0\rangle$ Burgers vectors associated with the glide dislocations on the basal plane. This mode of slip constitutes only two independent modes since the sum of any two slip vectors is equal to the third. Thus, $AB + BC = AC$.

Slip is also known to occur in beryllium on the three primary prism planes designated $ABB'A'$, $BCC'B'$, and $ACC'A'$. The Burgers vectors of the dislocations which are involved in prism slip are also the AB, BC, and CA vectors. The three additional modes namely $(ABB'A')$ $[AB]$, $(BCC'B')$ $[BC]$, and $(ACC'A')$ $[CA]$ represent two additional independent modes of deformation.

II. EXPERIMENTAL

Single crystals of beryllium were grown from the melt with the desired orientation using a seeding and floating zone melting technique developed and used by Spangler et al. [6]. The crystals receive about 8 to 12 passes in zone refining. The crystals "as grown" were rough machined into tensile bars using spark discharge techniques. They were then electropolished to remove the surface layers deformed by spark machining. The tensile specimens had gauge sections $\frac{5}{8}$ in. long and approximately $\frac{1}{10}$ in. in diameter with conical shaped shoulders at the extremities for gripping (inset Fig. 2). The specimens were tested in an Instron tensile machine at a strain rate of $7 \cdot 10^{-4}$/sec, using two short lengths of steel chain at each end of the tensile grips to minimize nonaxiality during loading (Fig. 2). Two orientations examined are shown in Fig. 3. In Fig. 3a, the crystal is oriented so that the basal plane was approximately 45° to the tension axis and the projection of the tension axis on the basal plane was approximately parallel to a slip direction. In Fig. 3b, the crystal was oriented with the basal plane parallel to the tension axis. Slip usually commenced on a single prism plane. At large strains, duplex slip occurred with two cooperating prism planes participating in the slip.

Crystals were deformed to various levels of strain, they were then removed from the straining apparatus and examined with the optical microscope for slip lines. Replicas were also made of the surface and these were examined in the electron microscope. The technique used for replicating beryllium crystals was that used by Grube and Rouze [7], in which a water soluble agent Victawet 35B was first evaporated on the surface of interest followed by the evaporation of a SiO layer. The replicas were easily removed by immersing the specimen in distilled water. The replicas floated on the water and were readily lifted onto grids and shadowed with tungsten oxide. Details of the slip lines were observed by examination of the replicas in the electron microscope.

The technique used for preparing foils from bulk beryllium single crystals for electron transmission microscopy was devised by Wilsdorf and Wilhelm [8]. After the surface layers were thoroughly examined using replicating techniques, the specimens were sectioned using spark cutting techniques so that the plane of the section was closely parallel to basal plane. Depressions were placed on adjacent sides of the sections so that the thickness of

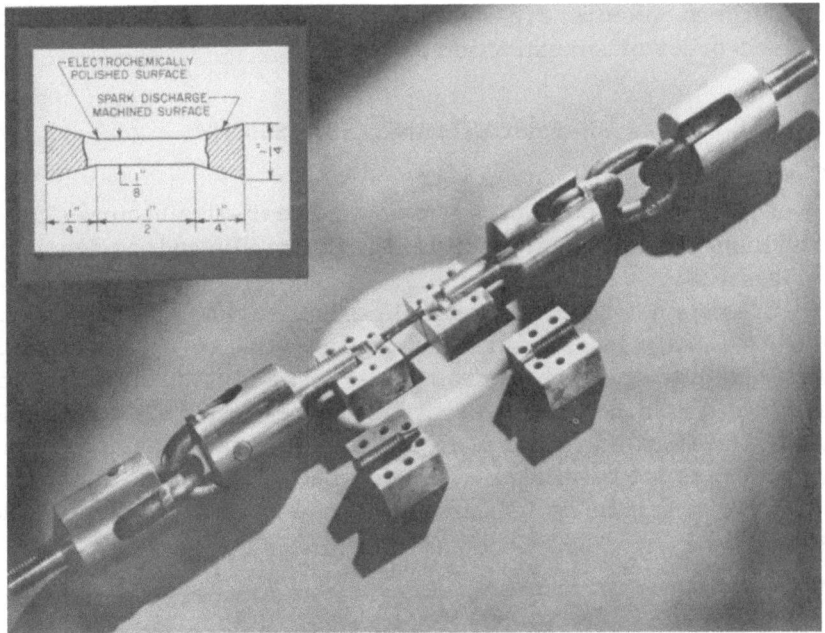

Fig. 2. Single-crystal tensile specimen, tensile grips, and alignment chains.

metal at the base of the depression was approximately 0.05 mm. The specimens were then given a final electropolish in a solution containing 90 parts phosphoric acid, 30 parts sulphuric acid, 30 parts ethanol, and 30 parts glycerine. The polishing was interrupted when a small hole just appeared at the base of the depression. Areas adjacent to the hole were generally found to be transparent to electrons. Dislocations having either *AB*, *BC*, or

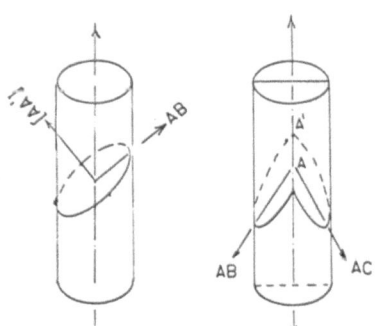

Fig. 3.(a) Orientation of single crystal for basal slip, and (b) Orientation of crystal for prism slip.

CA Burgers vectors are visible in contrast at first–order and second–order prism extinction contours.

III. RESULTS AND DISCUSSION

Tensile Tests

Typical resolved shear–stress vs. shear–strain curves for a beryllium single crystal deformed for prism slip and one deformed for basal slip is shown in Fig. 4.

Crystals deformed for basal slip exhibited two stages of hardening only. Stage I was described as an easy glide region and stage II was a region of rapid hardening just prior to failure. Beryllium single crystals deformed for prism slip exhibited a macroscopic yield stress of the order of 15 times larger than the macroscopic yield stress for basal flow. Three distinct stages of work hardening were observed as indicated in Fig. 4 by stage I, stage II, and stage III.

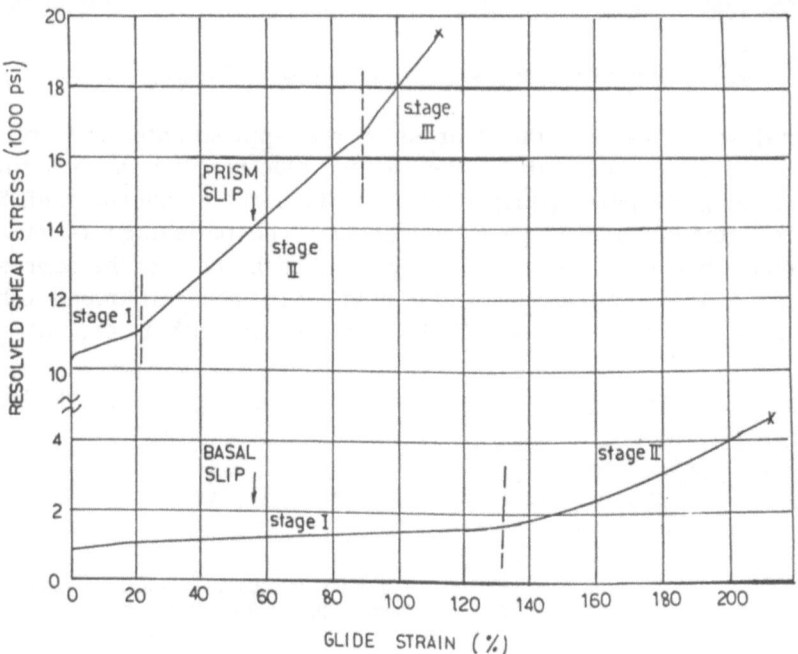

Fig. 4. Resolved shear–stress vs. glide strain for single crystals deformed in tension for basal glide and prism glide.

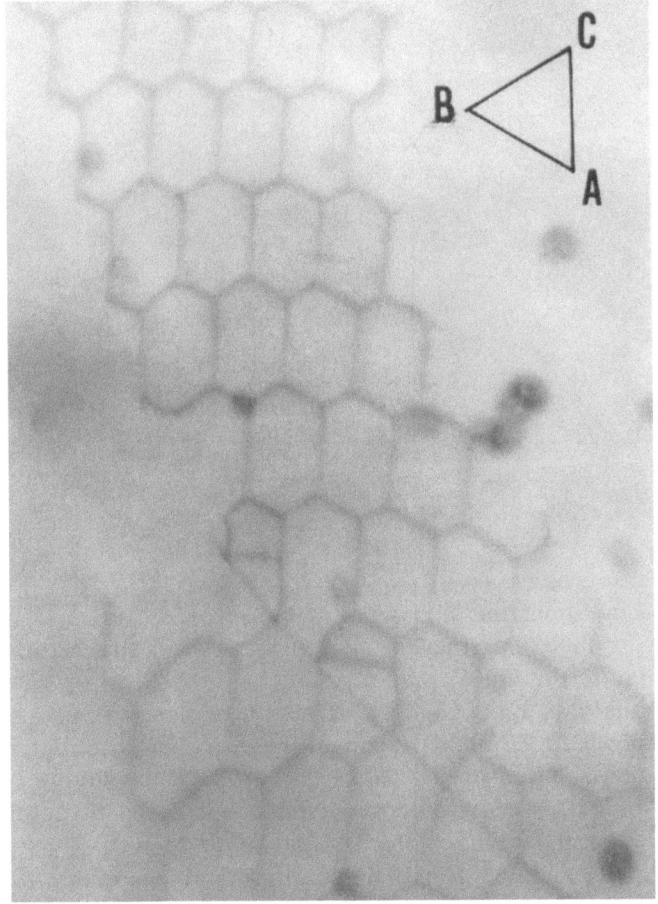

Fig. 5. Hexagonal network of dislocations observed in as-grown
high-purity beryllium crystal.

Dislocations in As-Grown Crystals

A study of the dislocations in foils cut from as-grown crystals
revealed that most of the dislocations were arranged in networks
and subboundaries. Hexagonal networks as shown in Fig. 5 ranged
in size from a mesh of 0.1 to $1\,\mu$. The nets were planar and lay in
the (0001) plane. They were made up of dislocations having $\frac{1}{3}$
$\langle 11\bar{2}0 \rangle$ Burgers vectors. This was readily determined by observing
the contrast at $(10\bar{1}0)$ extinction contours. The dislocations having a
Burgers vector lying in the plane contributing to the diffraction

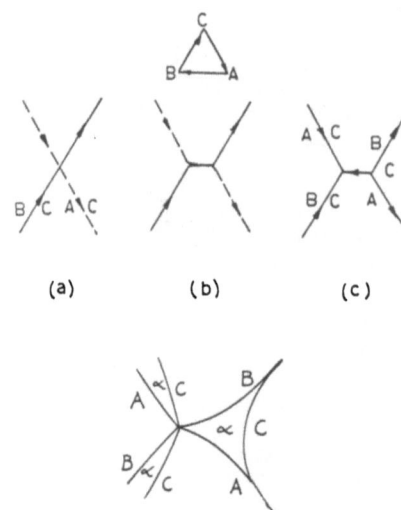

Fig. 6. (a) Two intersection dislocations with different Burgers vectors on (*ABC*) plane. (b) Interaction occurs at point of intersection. (c) Interaction produces threefold nodes. (d) Low-stacking fault material exhibits node extension.

exhibited no contrast and were not visible. The nets showed no contrast at (0002) extinction contours, proving that their Burgers vectors lay in the (0001) plane.

The analysis of hexagonal nets with coplanar Burgers vectors is well known. If Frank's notation is used, the interaction occurring between two intersecting screw dislocations having Burgers vectors *AB* and *BC* results in the formation of the threefold nodes consisting of the three dislocations *AB*, *BC*, and *CA* as shown in Figs. 6a, 6b, and 6c. For high-stacking fault materials, the nodes in the nets are expected to be contracted as shown in Fig. 6c. Low-stacking fault energy materials exhibit an extension of the nodes as shown in Fig. 6d.

The nodes in the nets shown in Fig. 6 are seen to be contracted, suggesting that the stacking fault energy of high-purity beryllium is high. The presence of small quantities of iron, however, is believed to have an appreciable effect on the stacking fault energy, and networks observed in these materials were reported [9] to have exhibited these extended nodes. This suggested that the stacking fault energy could be reduced by alloying.

One would expect that high-purity beryllium with a high-stacking fault energy would thus exhibit easy cross slip from the (0001) to (10$\bar{1}$0) planes as previously described.

Basal Plane Slip

Examination of foils cut from high-purity beryllium single crystals deformed in stage I showed the presence of straight-edge pairs and edge dipoles as shown in Fig. 7. These observations suggested that screw dislocations gliding on the basal plane in the easy glide region have considerably more mobility than edge dislocations and readily glide out of the crystal leaving edge dislocations. Numerous mechanisms have been proposed in the literature which attempt to explain the formation of edge dipoles [10−12].

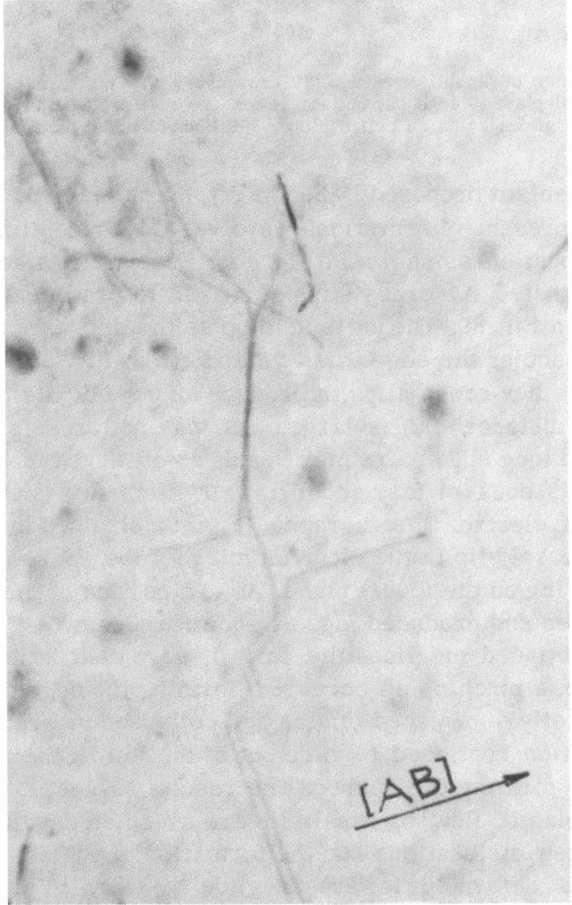

Fig. 7. Pairs of long straight-edge dislocations observed in foils cut from bulk crystal deformed by basal glide. Magnification 40,000×.

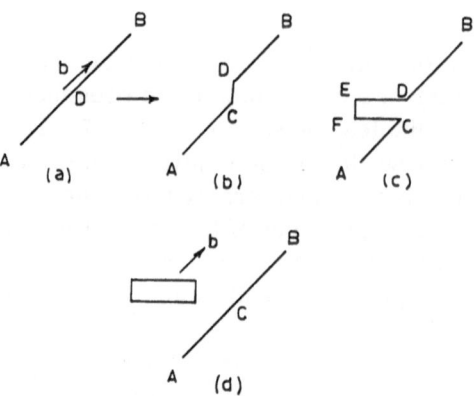

Fig. 8. Mechanism of dipole formation. (a) Screw cross slips on prism plane. (b) Screw glides on basal plane — edge part of jog is held up. (c) Edge dipole results when dislocation cross glides again on prism plane. (d) Loop forms by pinching off edge pair.

A mechanism proposed by Segall [13], which seems to be applicable to the present observations, involves the cross slipping of the screw dislocation as shown in Fig. 8a. Essentially a screw dislocation designated AB cross slips along AD to produce a larger jog DC shown in Fig. 8b. The jog is held up as the dislocation continues to glide producing the edge pairs DE and CF in Fig. 8c. The screw portion DB then cross slips to produce an edge dipole loop. If the cross slip distance DC is large, DB may not cross slip to close the loop and long edge pairs may result as observed in Fig. 7.

That this does in fact occur in beryllium foil is shown in the sequence of electron micrographs (Fig. 9) obtained by deforming a foil for basal slip in the electron microscope. Screws were observed moving on the (0001) plane. At A, a portion of the screw had cross slipped and produced jogs as shown in Frame a. As the dislocation continued to glide, the pair of edge dislocations formed. In Frame b, a pinching off occurred, forming the edge dipole. The loop eventually diminished in size and collapsed in Frame c, while the dislocation continued to glide out of the foil. Other edge pairs can be seen forming in Frame c. The results suggest that the early stages of plastic flow by basal slip are associated with the movement of screw dislocations and the formation of edge dipoles.

Crystals deformed for basal glide to stage II, the region of rapid work hardening, just prior to failure revealed the presence of numerous edge dislocations. The edge dislocations were arranged in groups or packets, as shown in Fig. 10. One feels that as

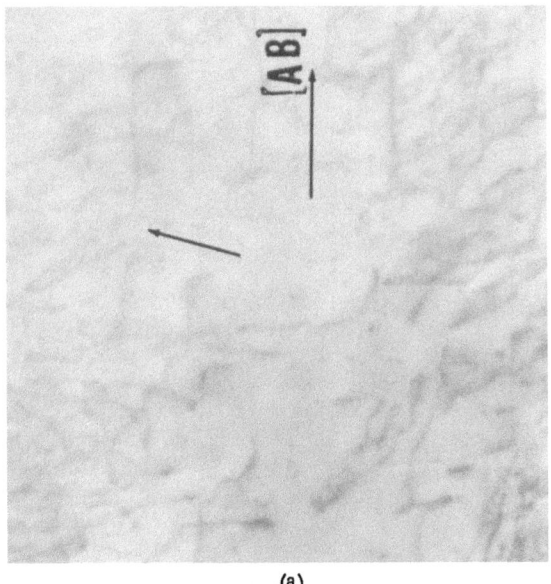

(a)

Fig. 9. Formation of edge dipole observed in foils deformed
in electron microscope. (a) Jogged screw dislocation held
up at A.

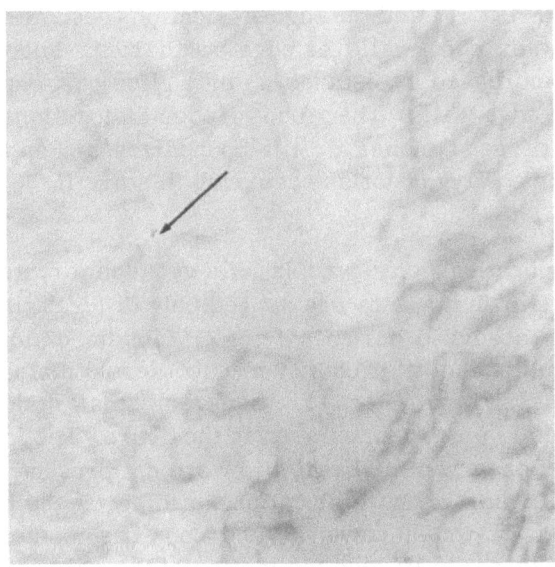

(b)

Fig. 9. (b) Pinching off results in edge dipole.

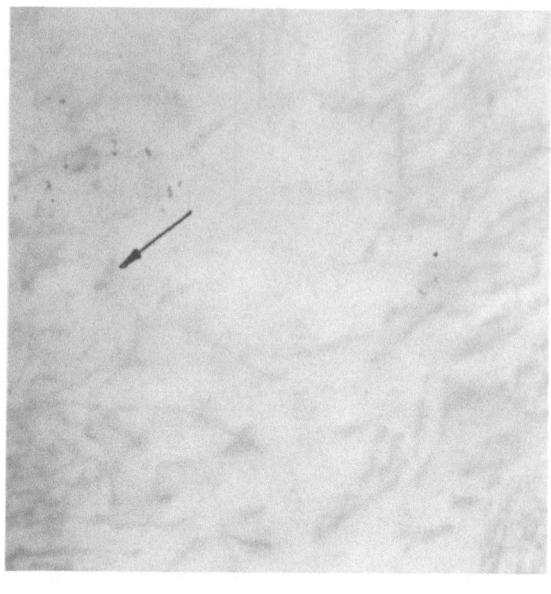

(c)

Fig. 9.(c) Collapse of loop occurs.

the density of edge dislocations increases, interaction between the gliding edges occurs and the edge dislocations become arranged into stable complex edge walls or edge boundaries. Edge dislocations encounter increased resistance as they glide past one another and through the edge walls. The groups of edge dislocations are similar in aspect to bend planes or tilt boundaries and appear to be the result of glide polygonization occurring in stage II.

Prism Slip

The dislocations observed in foils cut from crystals deformed for prism slip in stage I were appreciably different in appearance from those observed for basal glide. Prism dislocations were very irregular and a predominance of screw dislocations were observed, as shown in Fig. 11. Numerous small dislocation loops were frequently observed, as seen in the figure. The long straight-edge dipoles observed for basal slip were not present.

It is felt that screw dislocations gliding on the prism planes intersect the grown-in networks and soon become heavily jogged. The jogs then act to impede the motion of the screws. The theory is analogous to that proposed by Hirsch [14] for the hardening of all face-centered cubic metals in stage II. Hirsch proposed that screw dislocations become jogged when they intersect forest dislocations with Burgers vectors out of the slip plane. When the jogs move with the dislocation, both vacancies and interstitials may form, depending upon the nature of the jog. Vacancy jogs which move with the formation of vacancies lead to a resistance of movement. Interstitial jogs move conservatively since interstitials are believed to move with the dislocations and do not hold up the dislocations except at very low temperatures. Vacancies produced by the move-

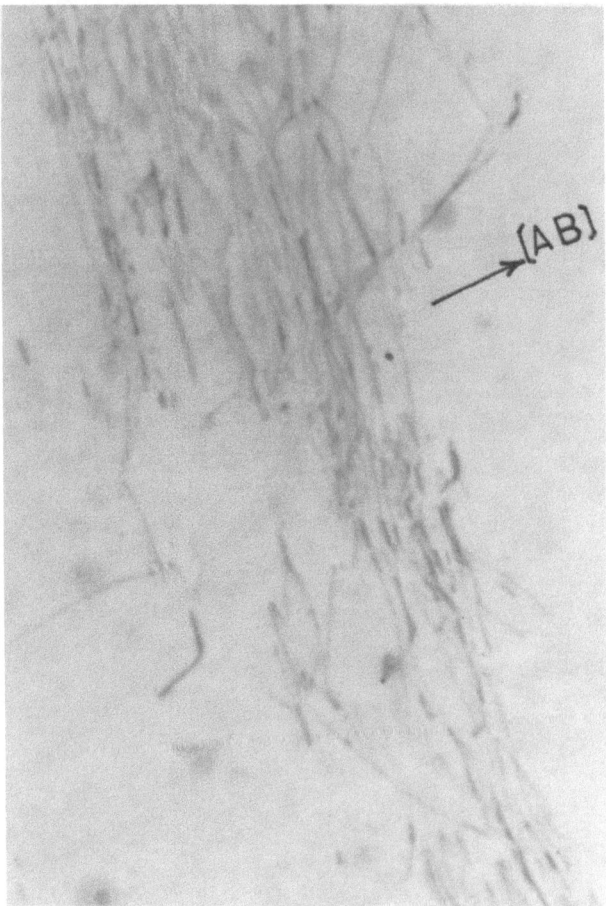

Fig. 10. Bundles of edge pairs observed in heavily deformed crystal — deformation by basal glide.

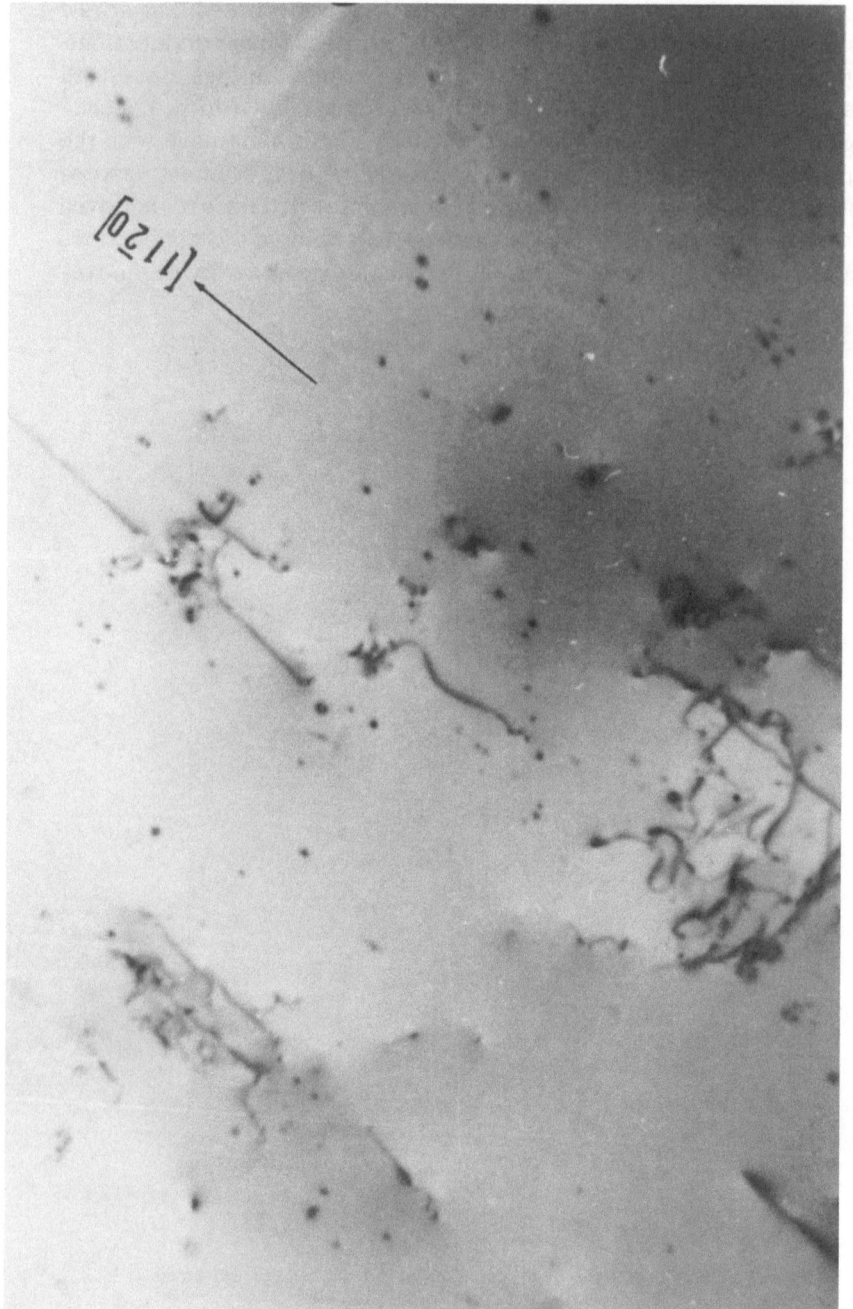

Fig. 11. Straight screw dislocations observed in foils deformed in stage I for prism slip. View on $(10\bar{1}0)$ slip plane. Magnification 40,000×.

ment of vacancy jogs may then lead to the formation of the dislocation loops observed in crystals deformed for prism slip. Edge dislocations gliding on the prism plane are not impeded by network intersections and these glide out of the crystal. Evidence to support this hypothesis was obtained by the observation of short slip lines appearing on the surface of a deformed bulk crystal deformed for prism slip and shown in Fig. 12. The crystal had been deformed approximately 10% in tension in a direction parallel to the basal plane. Examination of replicas of the slip lines with the electron microscope revealed, as seen in Fig. 13, that the broad bands were composed of fine slip lines which terminated in the surface. The ends of the slip lines were connected by slip lines parallel to the basal trace.

These observations suggested that the screw dislocations held up by jogs in the prism plane eventually cross-slipped on the basal plane, producing the basal slip lines observed in Fig. 13.

No detailed information regarding the hardening in stage II for prism slip has been obtained at present. A foil cut from one crystal deformed to this level of strain exhibited complex tangles of dislocations, as shown in Fig. 14.

In stage III a second prism slip system begins to operate. An interaction between edge dislocations of the two systems is believed to occur along the line of intersection of the two prism planes as shown in Fig. 15. Edge dislocations with AB Burgers vectors gliding on $ABB'A'$ interact with edge dislocations having BC Burgers vector on $BCC'B'$. The result of this interaction is the formation of dislocations lying along the line of intersection BB' and having Burgers vector CA. Since the applied stress is along Aa, the Burgers vector of the dislocation formed lies on a plane normal to the applied stress, and the boundary formed is stable and cannot glide. These boundaries were evidenced in crystals deformed in stage III as shown in Fig. 16.

IV. SUMMARY

1. Bulk high-purity beryllium single crystals were deformed in tension for basal and for prism slip.

2. The stress–strain curves were compared, revealing a macroscopic yield stress for prism slip of the order of 15 times that for basal slip. The work-hardening rate for prism slip was observed to be greater than that for basal slip.

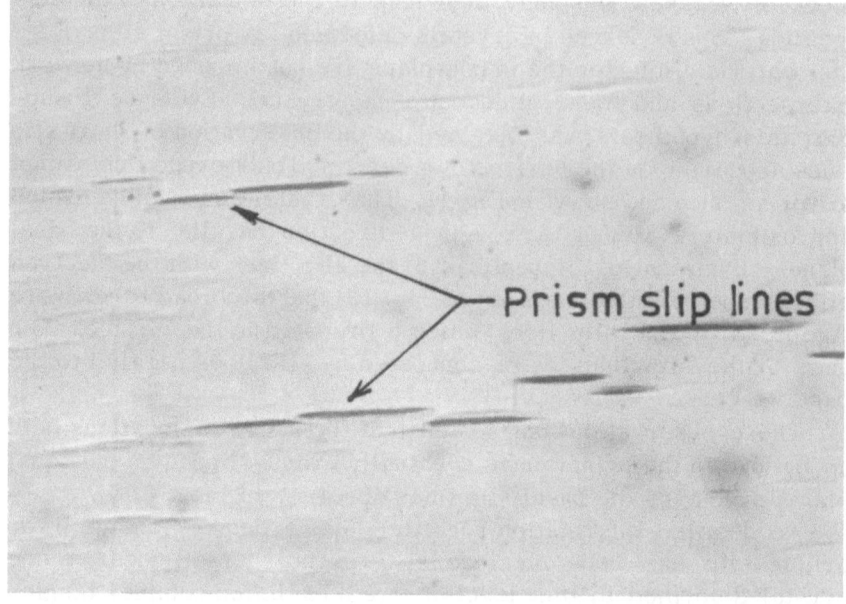

Fig. 12. Short prism slip lines observed on surface of crystal deformed for prism slip in stage I. Magnification 690×, optical microscope.

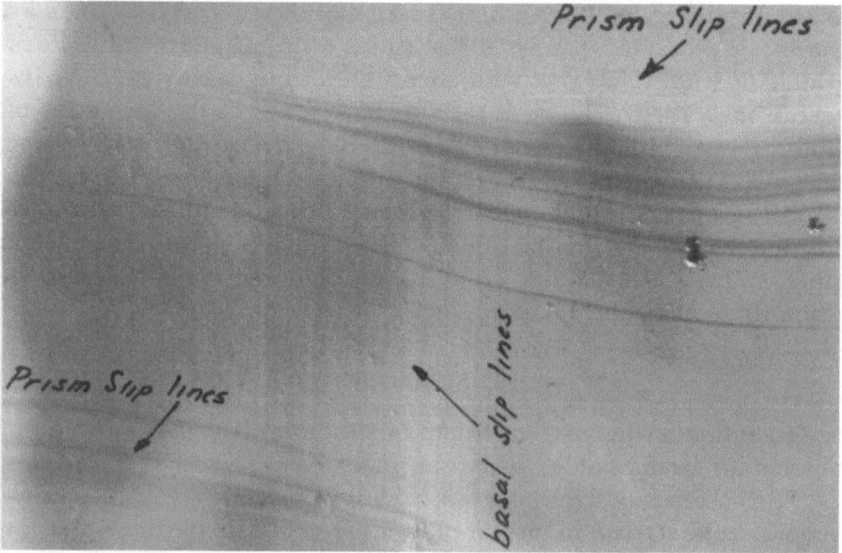

Fig. 13. Broad bands of prism slip composed of finer slip lines. Basal slip lines connect the ends of prism slip lines. Magnification 15,000×, reduced 20% for reproduction.

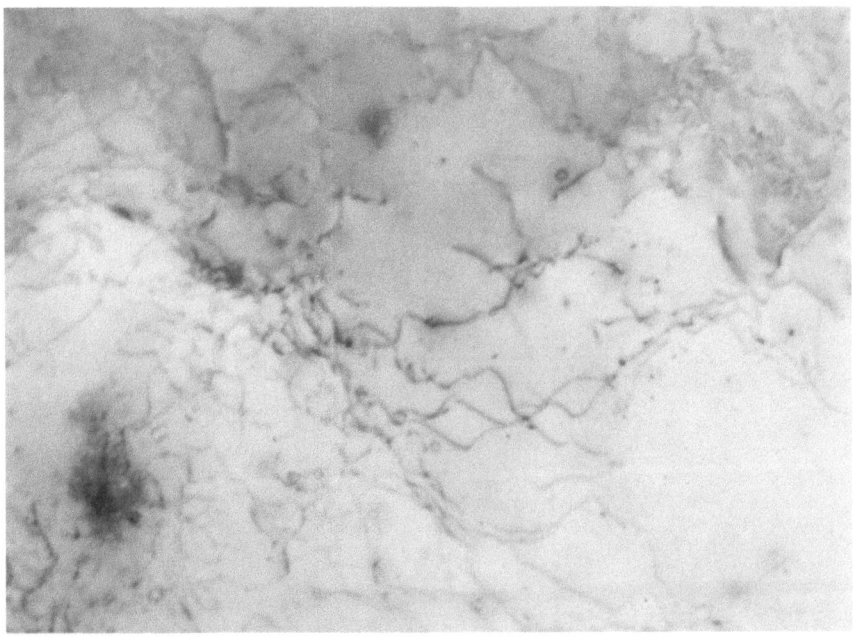

Fig. 14. Dislocations observed in stage II of prism slip arranged in complex tangles. View on (10$\bar{1}$0) slip plane. Magnification 40,000×.

3. Examination of dislocation networks in as-grown crystals reveals only Burgers vectors of $\frac{1}{3}$ $\langle 11\bar{2}0 \rangle$ type, and those which were observed lay in the basal plane.

4. Examination of foils cut from the bulk crystals deformed in stage I for basal slip revealed straight-edge dislocation pairs and elongated edge dipoles.

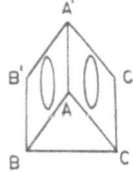

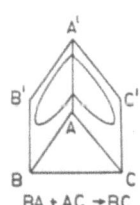

Fig. 15. Glide dislocation on $ABB'A'$ with Burgers vector AB interacts with glide dislocation on $ACC'A'$ having AC Burgers vector to produce dislocations along line AA' with Burgers vector BC.

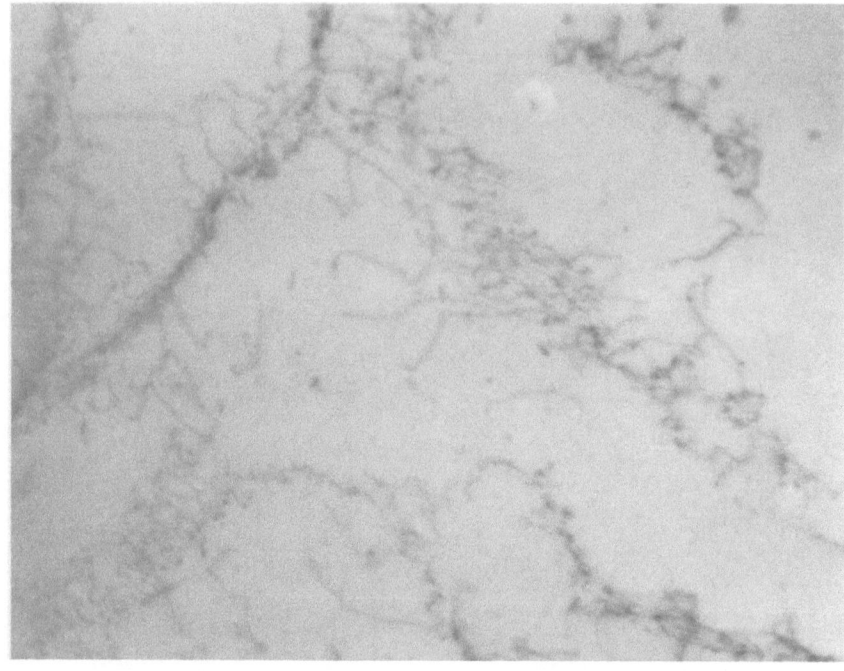

Fig. 16. *BC*-type boundaries viewed in foil cut from bulk crystal deformed to stage III, for prism duplex slip. View on (0001). Magnification 40,000×.

5. At high strains in stage II for basal slip, bundles of edge dislocations were observed.

6. Examination of foils cut from bulk crystals deformed to stage I for prism slip revealed a predominance of screw dislocations.   In addition, slip line studies revealed the presence of short prism-slip-line lengths of emerging edge dislocations. Numerous small dislocation loops were also observed in the foils cut from these crystals.

7. The onset of duplex prism slip in stage III is associated with a further increase in the hardening rate.   Foils cut from these crystals exhibit subboundaries not observed in crystals deformed in stage I and stage II.

V. CONCLUSIONS

1. Stage I hardening for prism and basal slip is associated with the intersection of the glide dislocations with the grown-in net-

works. Since the Burgers vectors of the grown-in networks lie in the basal plane, intersections of the prism screw dislocations with the networks can produce jogs in the gliding prism screw dislocations but not in the gliding basal screw dislocations. This difference was believed to account for the more rapid rate of work hardening in stage I for prism slip than for basal slip.

2. Stage II hardening for basal slip is associated with the formation of edge dislocation boundaries. Complex tangles of dislocations are associated with stage II hardening for prism slip.

3. Stage III hardening for prism slip is associated with the onset of duplex prism slip where stable boundaries were observed. These boundaries were felt to arise from the interaction of dislocations on intersecting prism planes.

4. No stage III hardening was observed for basal slip. The crystal failed in stage II.

VI. ACKNOWLEDGMENT

The authors wish to acknowledge the Department of the Navy, Bureau of Naval Weapons for the support of this work. They would also like to express their appreciation to G. E. Spangler for his help and assistance in the preparation and tensile testing of single crystals.

REFERENCES

1. J. J. Gilman, Trans. AIME 221:456 (1961).
2. P. Ward Flynn, J. Mote, and J. E. Dorn, Trans. AIME 221: No. 6, 1148 (1961).
3. A. T. Churchman, Proc. Roy. Soc. A226:216 (1954).
4. N. Thompson, Proc. Phys. Soc. 66:481 (1953).
5. V. V. Damiano, Trans. AIME 222:788 (1963).
6. G. E. Spangler, M. Herman, and E. J. Arndt, Franklin Institute, Final Report F-A2476, Department of the Navy, Bureau of Naval Weapons, Contract No. NoW 61-0221-d.
7. W. L. Grube and S. R. Rouze, Proc. ASTM 52:573 (1952).
8. F. Wilhelm and H. G. F. Wilsdorf, Franklin Institute Report on Air Force Contract No. AF 33(616)7065 (1961).
9. G. V. T. Ranzetta and V. D. Scott, J. Nucl. Mater. 10:113 (1963).
10. W. G. Johnston and J. J. Gilman, J. Appl. Phys. 31:632 (1960).
11. A. S. Tetelman, Acta Met. 10, 813 (1962).
12. P. B. Price, Phil. Mag. 5, 873 (1960).
13. L. Segall, Electron Microscopy and Strength of Crystals (Interscience, New York, 1963), p. 515.
14. P. B. Hirsch, Symposium on Internal Stress and Fatigue in Metals, General Moton Laboratories (1958).

The Use of Changes in X-Ray Diffraction Line Broadening to Study Recovery Kinetics in Pure Cobalt

C. R. Houska

Virginia Polytechnic Institute, Blacksburg, Virginia

X-ray diffraction line profile techniques have become sufficiently refined to give a detailed description of cold-worked structures. Local strain, small particle size, and stacking faults cause a broadening of the diffraction lines. Some profiles are broadened by all three, while others are influenced only by local strain and small particle size. These techniques were used to investigate recovery phenomena in samples of hexagonal cobalt.

Powder specimens were prepared by grinding a cobalt rod. This left each specimen in a highly cold-worked state. Separate specimens were taken to a series of temperatures in the recovery range, and the rate of sharpening at temperature of two diffraction lines was observed. One line initially was broadened by local strains and small particle size, while the second was broadened by all three effects, i.e., strain, particle size, and stacking faults. In each case, the initial rate of sharpening resulting from an increase in crystalline perfection could be related to a single activation energy. It is possible in principle to examine recovery kinetics in several crystallographic directions, and approximately to separate the changes resulting from a reduction in local strain and an increase in average particle size.

I. INTRODUCTION

Pure cobalt transforms on cooling from fcc to the hexagonal phase by a simple shearing of the close-packed (111) planes. This martensitic transformation has evoked considerable interest, mainly because of the simplicity of the shearing mechanism. Thermodynamic equilibrium occurs at 417°C, but appreciable hysteresis is observed between the heating (430°C) and cooling (390°C) transformations [1]. The kinetics of these reactions are predominantly athermal, although isothermal components are operative to some degree. These characteristics, together with the diffusionless nature of the phase change, permit one to designate it as martensitic. Some local transformation strains in the vicinity of a transformed platelet are to be expected. The rather high transformation temperatures would suggest that such transformation strains are likely to anneal out in seconds. The existence of resid-

111

ual transformation strains should have an important bearing upon
the shearing mechanism.   It was for this reason that experiments
were undertaken to examine recovery phenomena in pure cobalt [2].

II. DEFECTS IN HEXAGONAL COBALT AFTER
LONG-TERM ANNEALING AT LOW TEMPERATURES

X-ray measurements were conducted on powder briquettes prepared from cobalt of 99.99% purity.*   Powders were obtained by
grinding with an Alundum wheel and were separated from the wheel
residue with a magnet.   The -325 mesh powder was cold pressed
into 1 by $\frac{1}{2}$ by $\frac{1}{8}$ in. slabs and the resultant specimens were
completely hcp.   Pressing resulted in a slight preferred orientation
with the basal plane of the particles tending to lie parallel to the
compression surface.   The diffraction lines were broadened due to
the presence of local strain and small particle size.†

Starting with 100% hcp as-ground powders, a specimen was
annealed at 300°C for 2 hr.   The final structure was completely
hexagonal.   A local-strain and particle-size analysis of the (002)
and (004) reflections gave local strains that were very small (less
than $7 \cdot 10^{-4}$ based on a distance of 100 interplanar spacings) and an
average particle size in excess of 1000 A. These numbers indicate
that both strain and particle-size broadening are small giving relatively sharp (002) and (004) lines.   Hexagonal lines with indices
$h - k = 3t \pm 1,\ l \neq 0$ were found to be broadened due to the presence of
both deformation and growth stacking faults, which may be depicted
as follows:

Growth Fault	Deformation Fault
hcp hcp	hcp hcp
A B A B C B C	A B A B C A C A
fcc	fcc

The letters A, B, and C refer to the conventional stacking positions
of close-packed planes [4].   Either type of fault can be produced by
the growth of two out-of-phase hcp lattices together, while the deformation fault can also be formed by partial slip which converts A
planes into C planes and B planes into A planes.   A growth fault
contains three planes of fcc stacking, whereas the deformation
fault contains four.

*Obtained from Johnson, Matthey and Company, Ltd., London, England.
†A detailed description of the X-ray diffraction line shape analyses is given in reference 3.

TABLE I

Stacking Faults in Cold-Worked and Annealed hcp Cobalt

Annealing temperature, °C	Annealing time	Stacking fault probability $(1 : P)$*	
		Growth	Deformation
300	2 hours	1 : 31	1 : 53
300	7 days	1 : 26	1 : 125
300	7 days ⎫		
	⎬	1 : 28	1 : 250
390	2 days ⎭		

* P is the average number of reference planes sampled before one faulty sequence was discovered.

The density of both types of faulting is represented in Table I by means of the notation $1 : P$, where P is the average number of reference planes to be sampled before one faulty sequence is found. In general, growth faults were found to be more prevalent than deformation faults. Further annealing for a total of seven days at 300°C did not greatly change the probability of finding growth faults; however, the average spacing between deformation faults nearly doubled. A total anneal of seven days at 300°C and two days at 390°C again gave an increase in the average spacing between deformation faults to 1 in 250 with no significant change in the occurrence of growth faults. The shape of the X-ray lines could be explained on the basis of a random distribution of faults. It is evident that deformation faults can be removed by low-temperature annealing and that similar treatments have no effect upon the growth fault. It is likely that only recrystallization will remove stacking errors of this type.

The analyses which are required for this type of information are time consuming and highly specialized. However, they do give results which characterize the fine structure, and these probably cannot be obtained by other means. In this investigation, the more exact line shape analysis has been used to investigate the terminal structures which result from long-term annealing. A second approach was developed which permitted an investigation of the early stages of recovery. Of necessity, this must be both simple and fast. These advantages cannot be gained without making approximations and accepting a less rigorous approach.

III. RATE OF LINE SHARPENING USING INTEGRAL BREADTHS

A convenient measure of broadening from any cause is the integral line width which can be obtained by dividing the area under the curve of intensity vs. 2θ by the maximum height. If it is assumed that the individual broadening effects are additive,* one can obtain the following approximate equation for the integral breadth [5]:

$$B(hkl) = \frac{K\lambda}{L(hkl)\cos\theta} + E(hkl)\tan\theta + B_i(hkl) \tag{1}$$

where K is a constant, λ is the X-ray wavelength, $L(hkl)$† and $E(hkl)$ represent the average particle size and strain, respectively, along the normal to the planes (hkl), θ is the Bragg angle, and $B_i(hkl)$ is the integral breadth corresponding to the instrumental broadening. The additive form for the instrumental contribution is justified only if this term is relatively small. This is generally true for metallic filings which have not exhibited prior recovery. It is important to recognize the angular dependence of each term. For small θ, the particle-size term is weighted more strongly (approx. $1/\cos\theta$) than the strain term (approx. $\tan\theta$), and for high angles the weighting is reversed. In either case, the line sharpens with increasing particle size L and decreasing average strain E. The importance of each term depends upon the influence of the defect structure upon the (hkl) planes which are responsible for the diffraction line. Accordingly, directional effects can be investigated by selecting diffraction peaks from planes (hkl) of different orientation.

The initial rate of sharpening can be obtained by taking a time derivative of equation (1) with respect to L and E at $t = 0$.

$$\left(\frac{dB}{dt}\right)_0 = -\frac{K\lambda}{L_0^2\cos\theta}\left(\frac{dL}{dt}\right)_0 + \tan\theta\left(\frac{dE}{dt}\right)_0 \tag{2}$$

The two quantities, dL/dt and dE/dt, can be expressed in terms of separate rate equations

$$\left(\frac{dL}{dt}\right) = C_L\exp\left(-\upsilon_L/RT\right) \tag{3a}$$

---

*This result can be obtained if each contribution to the broadening has the shape of a Cauchy distribution.

†$L(hkl)$ will also be reduced by stacking faults if the particular line (hkl) is broadened by this defect.

$$-\left(\frac{dE}{dt}\right) = C_E \, \exp\left(-\upsilon_E/RT\right) \tag{3b}$$

where C_L and C_E are rate constants, while υ_L and υ_E are activation energies which depend upon the instantaneous defect structure. Since the fine structure changes with time, the activation energies are also expected to change. The present considerations have been limited to the initial rates of sharpening within heavily cold-worked materials. By substituting equation (3) into equation (2), we obtain

$$-\left(\frac{dB}{dt}\right)_0 = \frac{K\lambda \, C_L}{L_0^2 \cos\theta} \exp\left(-\upsilon_L^{\,0}/RT\right) + C_E \tan\theta \, \exp\left(-\upsilon_E^{\,0}/RT\right) \tag{4}$$

Figure 1 illustrates the rate of sharpening at 165, 200, and 230°C for the (002) line of hcp cobalt.* The specimens were prepared from grindings as described previously, and a freshly prepared sample was used at each temperature. The instrumental breadth $B_0(002) = 0.23$ (degrees 2θ) was subtracted from the total for each set of measurements, and the peak heights increased by a factor of about four as the value of B_0 was approached. Similar data were obtained using the (101) line, and this is given in Fig. 2. It was possible to measure both lines alternately with a single specimen held at the designated temperature. The integral breadth of the (101) line never reached the anticipated instrumental breadth because of residual broadening resulting from stacking faults (B_0

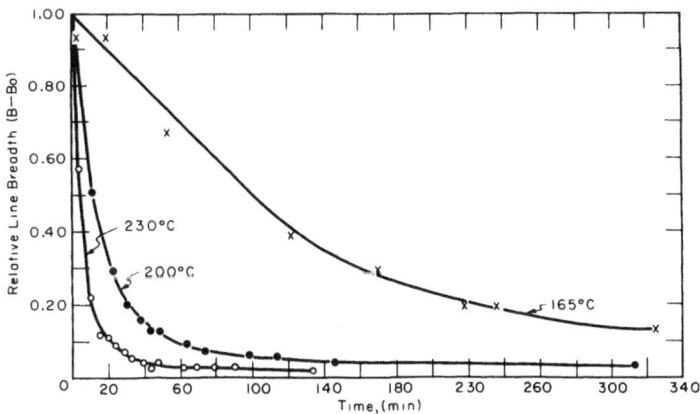

Fig. 1. Sharpening of (002) line for cold-worked hexagonal cobalt.

*The radiation used throughout these experiments was Fe K_α.

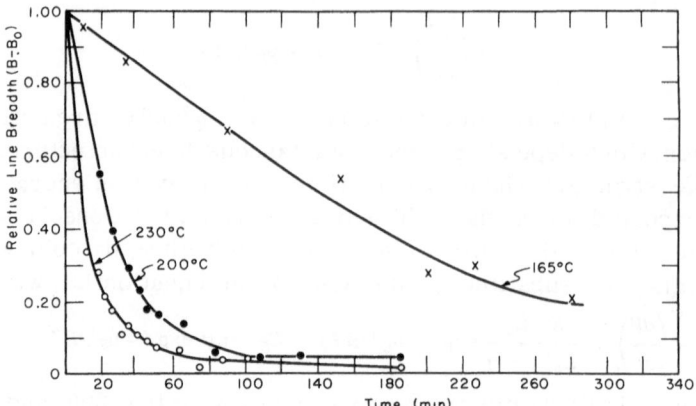

Fig. 2. Sharpening of (101) line for cold-worked hexagonal cobalt.

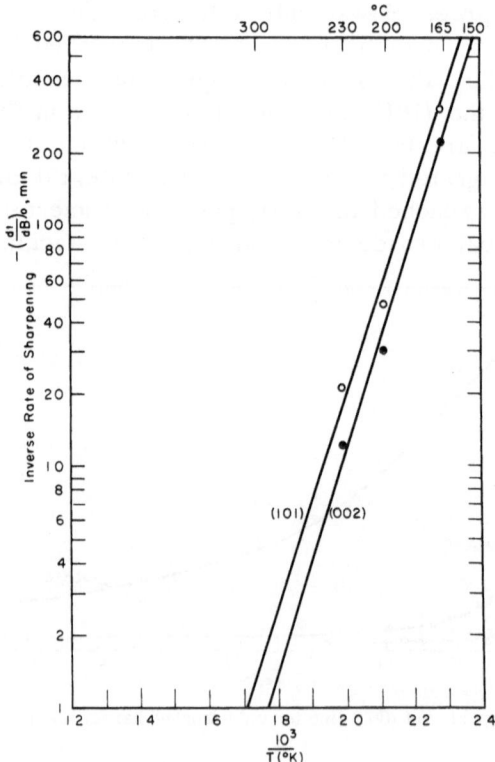

Fig. 3. Temperature dependence of inverse rate of sharpening.

(101) = 0.58 degrees 2θ). It was found, by another experiment, that the integral breadth of the (101) line was the same after seven days at 300°C as it was after seven days at 200°C.

Figure 3 is a plot of the initial value $\ln{(-dt/dB)_0}$ vs. $10^3/T$ (°K) for three temperatures and the two diffraction lines. It is evident that both sets of data have the same slope, but are displaced along the vertical axis. A reasonable fitting can be obtained by taking a single activation energy v^0 which yields

$$-\left(\frac{dB}{dt}\right)_0 = \left(\frac{K\lambda C_L}{L_0{}^2 \cos\theta} + C_E \tan\theta\right) \exp{(-v^0/RT)}$$

If the logarithm is taken, the equation becomes

$$\ln\left(-\frac{dt}{dB}\right)_0 = -\ln C' + \frac{v^0}{RT} \tag{4'}$$

where

$$C' = \frac{K\lambda C_L}{L_0{}^2 \cos\theta} + C_E \tan\theta$$

This form can be obtained if $v^0{}_L = v^0{}_E = v^0$, or if either dL/dt or dE/dt is a dominant term in equation (4). A choice cannot be made between these alternatives on the basis of the present data. It should be emphasized that the same effective activation energy has been obtained regardless of the orientation difference between the (002) and (101) planes and regardless of the fact that the (101) line is also broadened by stacking faults, while the (002) is only affected by local strains and small particle size.

Although a separation of the particle-size term (which may contain a contribution due to stacking faults) and the strain term was not attempted, it is possible to make this separation by an extrapolation to zero θ. This can be seen if equation (4) is multiplied by $\cos\theta$.

$$-\left(\frac{dB}{dt}\right)_0 \cos\theta = a + b\sin\theta \tag{5}$$

where

$$a = \frac{K\lambda C_L}{L_0{}^2} \exp\left(-\frac{v_L{}^0}{RT}\right)$$

and

$$b = C_E \exp\left(-\frac{v_E{}^0}{RT}\right)$$

If $B(T, t)$ has been measured for several orders (nh, nk, nl), then a plot of equation (5) against $\sin \theta$ extrapolates to a and the slope gives b. The activation energies v^0_L and v^0_E can be obtained by plotting the equations

$$\ln a(T) = \ln \frac{K \lambda C_L}{L_0^2} - \frac{v_L^0}{RT}$$

$$\ln b(T) = \ln C_E - \frac{v_E^0}{RT}$$

(6)

A similar extrapolation can be applied to equation (1), corrected for instrumental broadening, giving $L(T, t)$ and $E(T, t)$.

IV. DISCUSSION

Data obtained from the rate of sharpening of two X-ray diffraction lines provides direct evidence that local stresses are rapidly relieved above 300°C. At 300°C, 50% of the line broadening is lost within 1 min, while at 150°C, the required time is nearly 300 min. The activation energy for the recovery process was about 20,000 cal/mole. It has been established that most of the transformation range for pure cobalt lies above 300°C [1], in the region of rapid recovery. Consequently, transformation strains are expected to anneal out in seconds in the temperature range of extensive transformation.

The recovery data were obtained at relatively low values of 2θ where particle-size broadening is more heavily weighted than that resulting from strain. Warren [6] has shown that lattice tilt resulting from an edge dislocation can give a contribution to the particle-size broadening in addition to strain broadening. Thus, the surprisingly small particle sizes which are measured using line shape techniques can be interpreted as some sort of an average separation between dislocations. The annihilation of two edge dislocations during recovery would thereby result in two effects—an increase in the average particle size as well as a reduction in strain. The activation energy depends upon the mechanism which brings about the annihilation. It is not surprising, in view of this interpretation, to find a single activation energy for the initial recovery of a heavily cold-worked material. A time-dependent increase in the activation energy is to be expected as the heavy stress concentrations are progressively removed [4].

It has been found that long-term low-temperature annealing reduces the number of deformation faults, but there was found no similar change in the number of growth faults. Evidently, deformation faults can be removed more readily than growth faults; this may occur because the deformation fault can be "unslipped" merely by moving a partial dislocation across a close-packed plane. Also, the deformation fault contains four planes in the unstable fcc sequence, while only three are found for the growth fault. Consequently, one might expect the deformation fault to be of higher energy.

The defect structure after long-term anneals did not exhibit a significant amount of strain and particle-size broadening. A model fixing the partial dislocations associated with deformation faults within widely-spaced small-angle boundaries [7] would be in accord with these findings. However, agreement would also be found if the partials were situated within large-angle boundaries.

REFERENCES

1. C. R. Houska, B. L. Averbach, ard M. Cohen, Acta Met. 8:81 (1960).
2. C. R. Houska, Doctorate Thesis, "An X-Ray Investigation of the Structure and Transformation Characteristics of Cobalt," Metallurgy Dept., Massachusetts Institute of Technology, 1957.
3. C. R. Houska and B. L. Averbach, Acta Cryst. 11:139 (1958).
4. A. H. Cottrell, Dislocations and Plastic Flow in Crystals, Clarendon Press, Oxford, 1953, pp. 73 and 187.
5. C. S. Barrett, Structure of Metals, McGraw-Hill, New York, 1952, p. 428.
6. B. E. Warren, Progress in Metal Physics, Vol. 8, Pergamon Press, New York, 1959.
7. J. C. M. Li and B. Chalmers, Acta Met. 11:243 (1963).

Measurement of Applied Stress
by X-Ray Diffraction

H. M. Otte and A. L. Esquivel

Materials Research Laboratory, Martin Company, Orlando, Florida
(Appendix by W. E. Lauer*)

For the measurement of residual stresses, X-ray diffraction is potentially the best nondestructive method available. Experiments are described in which silicon bronze and α-brass samples were examined by X-rays while subjected to tensile deformation. The positions of the diffraction lines and their breadths were measured and analyzed. The special features of the experimental arrangement employed are presented and the method of analysis explained.

An important result of this study is the observation that the stress–strain curve obtained from the X-ray diffraction measurements follows the stress–strain curve obtained from a standard tensile test, contrary to what many previous investigators have reported with different experimental techniques. Furthermore, for materials which may form stacking faults upon deformation, the X-ray data must first be corrected for the effects produced from stacking faults before the stress is calculated.

I. INTRODUCTION

Residual stresses play an important role in the manufacture and quality-testing of metal products. There is good reason, therefore, for the metallurgical engineer to be interested in the measurement of residual strains or stresses. However, besides its importance in production techniques [1], stress measurement possesses its own relevance to basic scientific research [2]. Most crystalline defects have associated with them a local and sometimes long-range stress field. The interaction of these stress fields may govern, to a large extent, the physical and mechanical properties of solids. And insofar as they enter into the determination of the properties of solid materials, residual stresses are of importance to the research scientist and engineer.

II. METHOD OF ANALYSIS

X-ray diffraction does not directly measure residual stresses, but does directly measure residual strains. The principal effect

*Address: Martin—RIAS, Baltimore, Maryland.

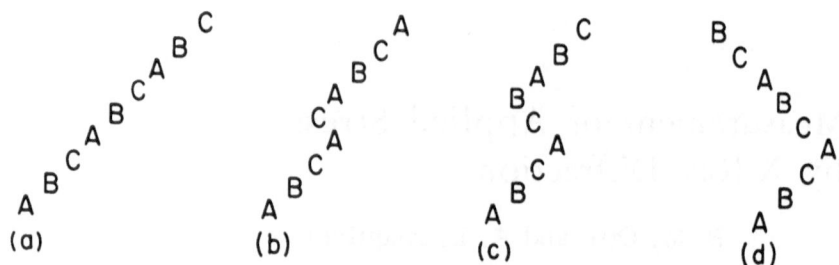

Fig. 1. Schematic diagram of stacking faults in fcc crystals: (a) normal fcc sequence, (b) deformation or intrinsic stacking fault, (c) extrinsic stacking fault, (d) twin stacking fault.

of residual stresses is the distortion of the crystalline lattice. Such a distortion will change the spacings of the atomic planes. From Bragg's law $\lambda = 2d \sin\theta$, it follows that a change Δd in the value of the interplanar spacing d, while keeping the X-ray wavelength λ constant, will alter or shift the scattering angle θ by $\Delta\theta$. Thus, differentiation of Bragg's equation yields $\Delta 2\theta = -2(\Delta d/d)\tan\theta$. If the strains are uniform throughout the specimen, the diffraction line will be shifted, but not broadened. On the other hand, if the polycrystalline specimen is subjected to an internal distribution of nonuniform stresses due to anisotropy and other factors, various values of the interplanar spacing d will arise such that the diffraction peaks will, in effect, be broadened and possibly also shifted. There are various methods of analyzing diffraction line broadening obtained from deformed metals [3-9]. However, in the present investigation which deals with the measurement of applied stresses, attention will be focused chiefly on the measurement of lattice parameter changes as observed from peak shifts.

The scale of the distortions introduced by the nonuniform stresses is important. If the distortions are essentially uniform over the distance of a coherently diffracting domain size, then they are termed macrostrains corresponding to macrostresses. If, on the other hand, the distortions are highly localized with respect to the size of the diffracting domains, then they are called microstrains. Macrostresses are essentially residual or applied stresses, whereas microstresses are associated with crystal defects. These crystal defects may also produce peak shifts and peak broadening.

The most important lattice defect encountered in our present work is a stacking fault [10,11]. Figure 1 shows a schematic representation of faults in the stacking sequence of {111} planes.

In Fig. 1a, the capital letters *A*, *B*, and *C* represent layers of atoms on the (111) plane of a normal fcc material. A stacking fault occurs when this regular sequence is disturbed. If the disturbance occurs only at one atomic layer, as in Fig. 1b, then the stacking fault is called a deformation or intrinsic stacking fault. If two successive layers are disturbed in their normal sequence, as in Fig. 1c, then a double deformation or extrinsic fault takes place. In Fig. 1d, a mirror image of the bottom layers is formed by the upper layers, and the fault at the boundary is called a twin fault.

Deformation faults are detected by peak shifts and symmetrical peak broadening in the diffraction patterns [11,12]. Double deformation or extrinsic faults are detected by means of both a nonsymmetrical broadening and a peak shift [13 – 15], while a twin fault [4,16] produces essentially only a nonsymmetrical peak broadening.

Our problem therefore consists of separating the effects due to residual stresses from those due to lattice defects, principally stacking faults. To solve this problem, a sufficient number of diffraction peaks must be obtained from the specimen. This may be achieved by using the Debye–Scherrer geometry together with the electronic counter methods offered by the diffractometer [17]. Such a diffractometer operated on a Debye–Scherrer arrangement presents a further advantage in that bulk specimens can be spun, thus providing a better sampling of the various types of reflections, resulting in an improvement of the accuracy of the lattice parameter measurements.

III. EXPERIMENTAL PROCEDURE

Measurements were made on two copper alloys: a polycrystalline Cu-30Zn α-brass with 29.75 at.%Zn, and a silicon bronze with 6.6 at.%Si–1.2 at.%Mn. The tensile specimens were prepared according to the standard ASTM specifications for a $\frac{1}{8}$-in. (3-mm) diameter test piece with a gauge length of $\frac{1}{2}$-in. (12 mm). After machining, the specimens were annealed in vacuum for 2 hr at 750°C and water-quenched.

Diffraction patterns were made with a General Electric XRD-5 diffractometer with its bisecting arm removed so that both halves ("positive" and "negative") of the Debye–Scherrer ring could be used. In Fig. 2 the general arrangement is shown with a specimen at (A) mounted on a spinner at (B). Cu K_{α_1} radiation, which was used throughout the investigation, was obtained by means of a

Fig. 2. Photograph of Debye—Scherrer diffractometer: (A) cylindrical specimen; (B) spinner motor; (C) X-ray tube; (D) monochromator and housing; (E) beam slit to cut-off K_{α_2}-component; (F) proportional counter; (G) diffractometer protractor without bisecting arm; (L) specimen platform.

Siemens curved quartz–crystal monochromator (D), Johansson focusing, and an adjustable beam slit (E) at the focal point of the monochromator to cut off the K_{α_2} component.

To obtain X-ray diffraction patterns of the tensile specimens while under uniaxial tension, a special holder (B, Fig. 3) was built and mounted on the platform (L, Fig. 3) affixed at the center of the diffractometer. This tensile specimen holder is described in greater detail in the Appendix. In order to obtain a representative sampling of reflections from the tensile specimens, a motor (I) was attached to the holder so that the specimen could be rotated at 170 rpm while being X-rayed under load or unloaded. The position of each diffraction line was determined reproducibly to $\pm 0.005°\ 2\theta$ from the center of gravity of the line profile. A final value of the line position was then obtained by averaging the readings from the positive and negative halves of the Debye–Scherrer diffractometer patterns.

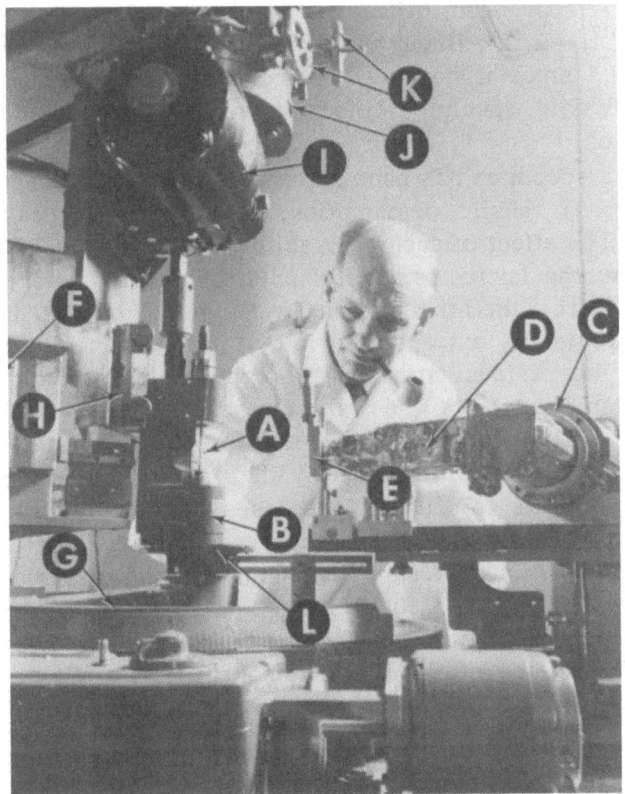

Fig. 3. Tensile specimen holder mounted on Debye–Scherrer diffractometer and connected to spinner motor: (A) cylindrical specimen; (B) tensile specimen holder; (C), (D), (E), (G), and (L) as in Fig. 2; (H) bracket with dovetail arrangement for tensile holder; (I) spinner motor for tensile specimen; (J) dual feed rotary table for centering specimen; (K) adjusting screws for rotary table.

IV. RESULTS AND DISCUSSION

The measured line positions were used to calculate apparent lattice parameter values. These values were then plotted against a suitable function that corrected for absorption effects and other systematic errors which varied with the scattering angle θ. The particular function used in the present work is originally due to Taylor and Sinclair [18] and Nelson and Riley [19]. Figure 4 shows how the apparent lattice parameter varies with this function, which is defined by $\frac{1}{2} [(\cos^2 \theta/\sin \theta) + (\cos^2\theta/\theta)]$. We chose the Nelson–Riley function for extrapolation because the apparent lattice parameter calculated from each diffraction peak varied linearly with this function [17]. Such a linear plot [20] is shown in Fig. 4a and labeled "unloaded," since it portrays lattice parameter values calculated from a tensile specimen before it has been subjected to tensile deformation.

After a specimen has been subjected to, or while a specimen is subject to, a tensile deformation, the diffraction peaks will be shifted. The effect of such peak shifts on the Nelson–Riley plot is to displace the lattice parameter points, indicated by open circles in Fig. 4a, from the straight-line fit designated by the dashed line. The change in the extrapolated lattice parameter value is used to calculate the residual strain; by using Hooke's law and the appropriate elastic constants, the residual stress can be determined. The numbers and subscripts shown refer to the sum of the squares of the Miller indices which identify the various reflecting planes.

If, in addition to residual stresses, stacking faults are also present in the specimen, it is observed that the slopes, $P_{3,12}$ and $P_{4,16}$, of lines drawn through first- and second-order reflections are changed (Fig. 4b). Such changes in slopes have been used to calculate the probability α of stacking faults in fcc specimens. The present figure, therefore, illustrates graphically the quantitative

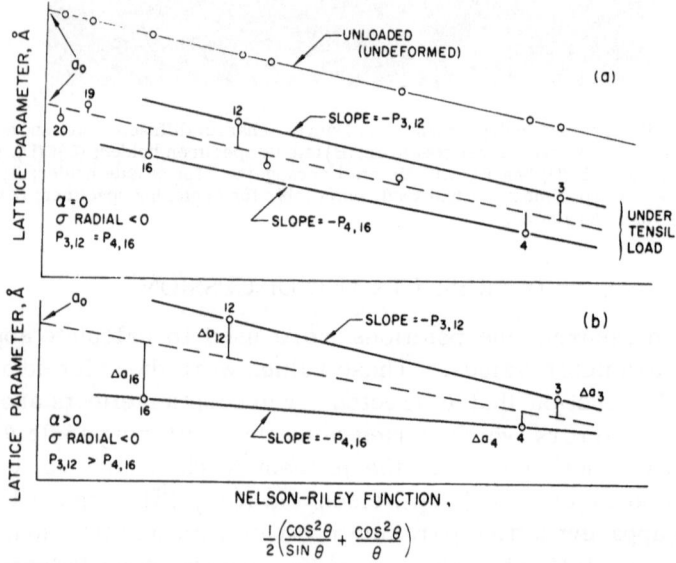

Fig. 4. Method of analyzing the data to obtain the extrapolated lattice parameter a_0 and the stacking-fault probability α. (From Otte & Welch, Phil. Mag., 1964.)

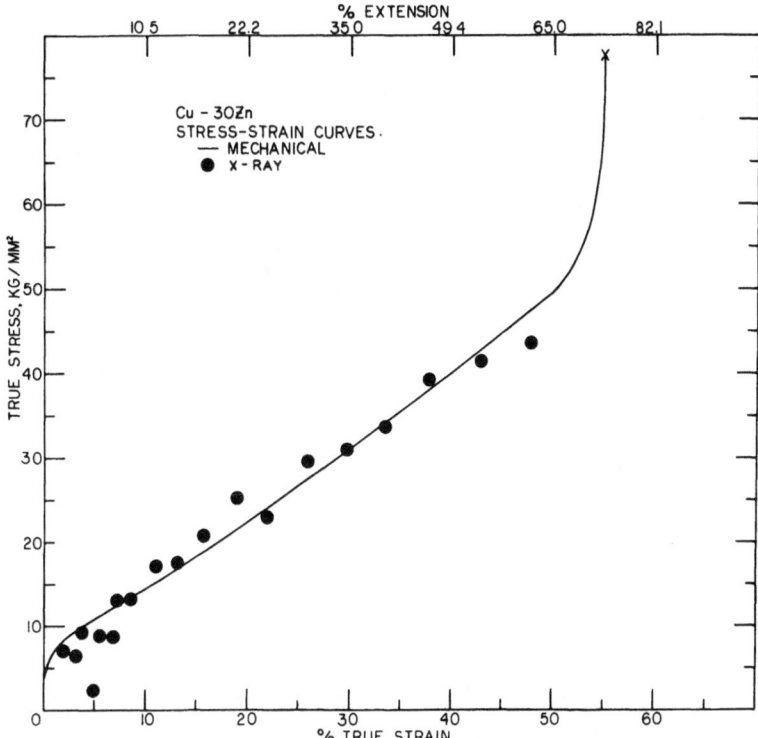

Fig. 5. Comparison of mechanical stress–strain curve (solid line) with X-ray stress–strain data (points) obtained from changes in lattice parameter for Cu–30Zn tensile specimen.

method of solving our problem of separating out the effects due to residual stresses and those effects due to lattice defects, such as stacking faults. From the extrapolated lattice parameter displacements, the stresses are determined and from the change in the slopes of lines drawn through selected orders of reflections, the stacking fault probability is calculated. Further details of the method have already been published and are available in the literature [20,21]. Only the results from some of our present work will be presented in this paper.

Values of the X-ray stress σ along the tension axis were calculated from the fractional change in spacing, or the lattice strain, $(\Delta a/a)$ perpendicular to the tension axis, using the extrapolated lattice parameter values obtained from an analysis of the data as in Fig. 4. The exact values of the stresses calculated are, however, somewhat

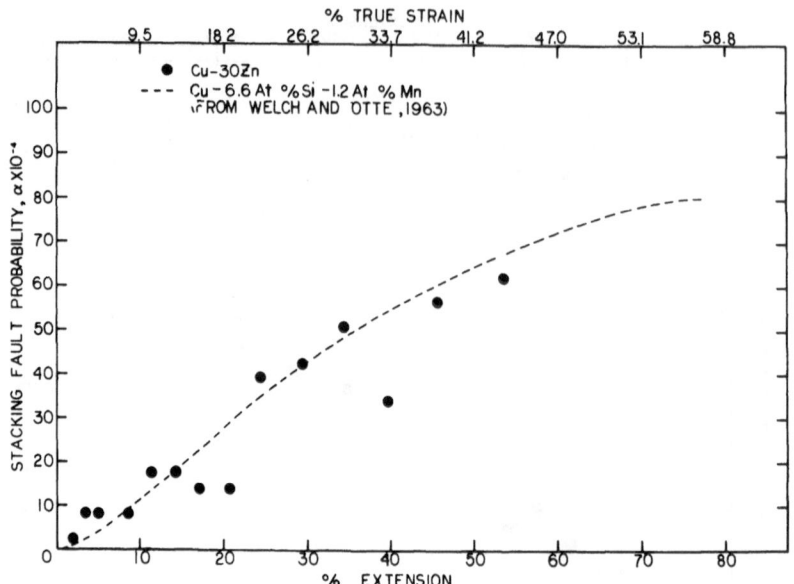

Fig. 6. Comparison of stacking-fault probability a for Cu–30Zn and Cu–6.6 at.%Si–1.2 at.%Mn specimens deformed in tension.

dependent on the elastic constants employed. The points on the graph in Fig. 5 correspond to values determined by the equation

$$\sigma = -(E/\nu)(\Delta a/a)$$
$$= -3.4\,(\Delta a/a) \cdot 10^4 \ \mathrm{kg/mm^2}$$

where E is Young's modulus and ν is Poisson's ratio. The value of $E/\nu = 3.4 \cdot 10^4 \ \mathrm{kg/mm^2}$ is slightly higher than that obtained by extrapolation of published data [22,23] for a-brass.

If the extrapolated elastic constant from published data is used, then the resulting points (Fig. 5) all fall slightly below the solid line of the mechanical stress–strain curve. This slight discrepancy between the mechanical and X-ray stress–strain curves may be a real effect suggesting a lower work hardening of the surface, although other explanations are possible. By using the extrapolated lattice parameters, effects due to stacking faults have essentially been corrected. However, because discrepancies between theoretically calculated and actually observed peak shifts due to stacking faults are known to exist, the extrapolated lattice parameters may not, in fact, be fully corrected for stacking-fault effects. At the moment, though, no further useful corrections are available.

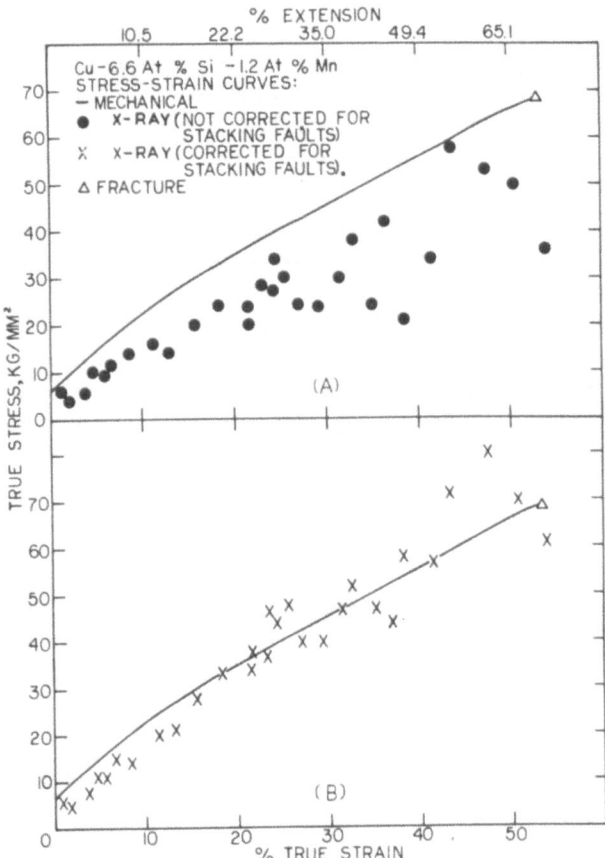

Fig. 7. Comparison of the mechanical stress–strain curve (solid line) of Cu–6.6 at.%Si–1.2 at.%Mn specimens with X-ray stress–strain data (points) obtained from displacements of (111)- and (200)-diffraction lines: (A) X-ray stress–strain data not corrected for stacking faults; (B) same data corrected for stacking faults.

The comparative magnitudes of the stacking-fault probabilities determined from the deformed α-brass and silicon bronze specimens are shown in Fig. 6. The stacking-fault probabilities were calculated from changes in the slopes of Nelson–Riley plots of the type shown in Fig. 4. From Fig. 6 it is seen that the stacking-fault probability increases with increasing tensile load, and that the probability of stacking faults in the α-brass specimen is comparable with that in the silicon bronze specimen studied in a previous investigation[20].

Thus, in the α-brass, the stresses calculated from the displacements of a pair of diffraction peaks differ measurably from those calculated by using the extrapolated lattice parameter. This was also true for the silicon bronze, and Fig. 7 illustrates the magnitude of the effect due to stacking faults. In Fig. 7a and b, the solid line represents the mechanical stress–strain curve for the silicon bronze tensile specimen. The points in Fig. 7a represent stress values which were calculated without correcting for the peak shifts due to stacking faults, whereas in calculating the stress values (crosses) in Fig. 7b, the proper corrections were made for diffraction effects due to stacking faults.

The stress values in Fig. 7 were calculated, from the displacements Δa of the lattice parameters from the straight line fit in the plot of the Nelson–Riley function vs. lattice parameter (Fig. 4). For lines 3 and 4 (corresponding to the (111) and (200) reflections,

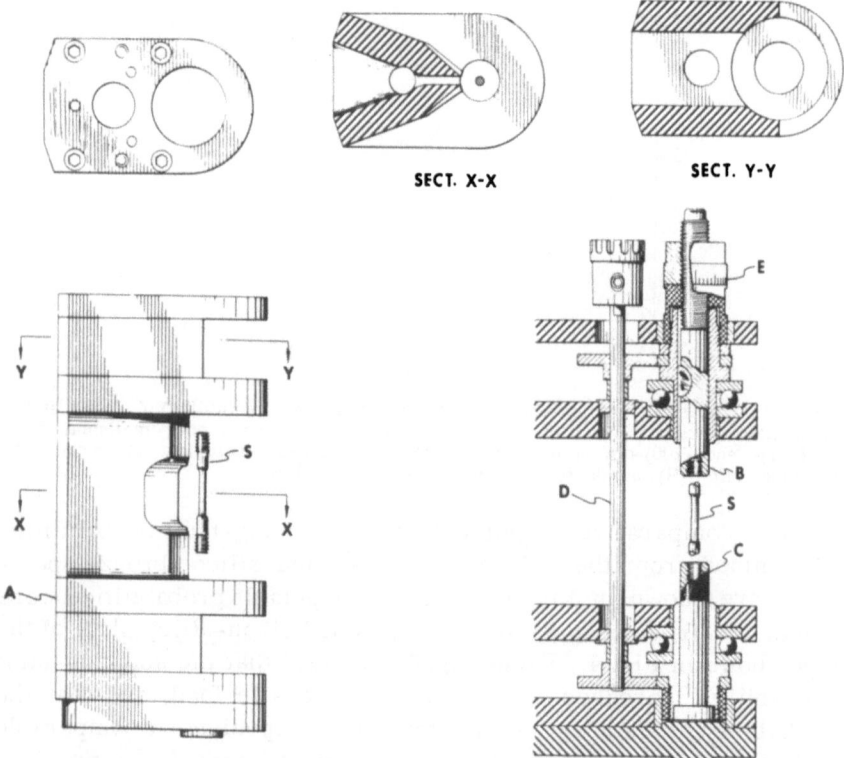

SECT. X-X   SECT. Y-Y

Fig. 8. Diagrammatic views of tensile specimen holder. (S) Tensile specimen; (A) steel frame of tensile specimen holder; (B) and (C) specimen grips; (D) transmission shaft for specimen rotation; (E) calibrated tension screws. (X-X) and (Y-Y) are cross sections of indicated areas.

respectively), the tensile stress σ along the specimen axis and the stacking fault probability α are given by

$$(\Delta a_4 - \Delta a_3)/a = (\overline{K}_4 - \overline{K}_3)\,\sigma + (G_4 - G_3)\,\alpha$$

The average values of the elastic constant K and the stacking-fault probability coefficients G have been tabulated elsewhere [20,21].

V. CONCLUSIONS

For both the polycrystalline copper alloys examined, it has been shown that the lattice stress–strain curve obtained by X-ray diffraction follows closely the mechanical stress–strain curve. When the stacking-fault probability becomes appreciable, as deformation of the specimen proceeds, corrections for effects due to stacking faults must be made when analyzing the peak shifts. Furthermore, such corrections must also be taken into account in cases where the diffraction peaks have suffered broadening and asymmetry. The necessity of taking into account the effect of stacking faults when measuring residual stresses in fcc materials that are readily prone to faulting has been clearly demonstrated in this paper.

Results similar to those presented and discussed here have also been obtained for other copper and silver alloys with low stacking-fault energies [24]. Although different sample geometries and experimental arrangements were employed in the other work, the basic conclusions of this paper were confirmed.

ACKNOWLEDGMENTS

The authors wish to thank the Office of Naval Research which supported this investigation, and Mr. Wallace Montague for his assistance in the experimental work. Acknowledgment is given to R. P. I. Adler and A. G. Crocker for their helpful comments.

VI. APPENDIX—THE TENSILE SPECIMEN HOLDER*

Figure 8 gives several sectional drawings of the tensile specimen holder (B, Fig. 3) used for the polycrystalline copper alloys described in the text. The salient parts of this holder are the tensile grips (B and C, Fig. 8) mounted at each end of the steel frame (A). These grips are geared to a shaft (D) so that they both rotate simultaneously without twisting the specimen (S). Tension

*By W. E. Lauer, Martin—RIAS, Baltimore, Maryland.

is applied by tightening a nut (E) which is threaded to the upper grip (B). This nut (E) is calibrated to indicate elongation in increments of 0.001 in., thus permitting calculation of the applied tensile strain.

The tensile specimen holder rests on a platform (L, Fig. 3) in the center of the X-ray diffractometer table and is held immobile by a bracket (H) supported from above. The tensile specimen holder is held to the bracket (H) by a dovetail arrangement to permit quick removal and accurate replacement of the holder. The bracket (H) also supports the motor (I) that provides the power for spinning the cylindrical specimen on its axis.

Precise alignment of the spinning specimen with the axis of the X-ray diffractometer (G) is obtained by use of a dual feed rotary table (J) with adjusting screws (K). The dual feed provides movement of the tensile specimen holder about a vertical axis thus keeping the tensile holder (B) clear of the diffracted X-ray beam at certain positions of the diffractometer table. The bracket (H) is bolted to the rotary table (J). The rotary table (J) and bracket (H) assembly are supported by an overhead structure.

REFERENCES

1. R. F. Thomson, "Engineering Interest in Internal Stresses," in: Internal Stresses and Fatigue in Metals, edited by G. M. Rassweiler and W. L. Grube, Elsevier Publishing Co., Amsterdam, 1959, pp. 3–14.
2. C. S. Barrett, "Scientific Interest in Internal Stresses," in: Internal Stresses and Fatigue in Metals, edited by G. M. Rassweiler and W. L. Grube, Elsevier Publishing Co., Amsterdam, 1959, pp. 15–40.
3. B. E. Warren and B. L. Averbach, J. Appl. Phys. 21:595 (1950).
4. M. S. Paterson, J. Appl. Phys. 23:805 (1952).
5. B. E. Warren and E. P. Warekois, Acta. Met. 3:473 (1955).
6. A. Fingerland, Czechoslov. J. Phys. 10B:233 (1960).
7. A. J. C. Wilson, Proc. Phys. Soc. 80:286 (1960).
8. A. J. C. Wilson, Proc. Phys. Soc. 81:41 (1963).
9. A. J. C. Wilson, Proc. Phys. Soc. 82:986 (1963).
10. B. E. Warren, Progr. in Metal Phys. 8:147 (1959).
11. C. N. J. Wagner, Acta Met. 5:427 (1957).
12. C. N. J. Wagner, Acta Met. 5:477 (1957).
13. C. A. Johnson, Acta Cryst. 16:490 (1963).
14. B. E. Warren, J. Appl. Phys. 34:1973 (1963).
15. H. M. Otte and H. Chessin, (to be published).
16. J. B. Cohen and C. N. J. Wagner, J. Appl. Phys. 33:2073 (1962).
17. H. M. Otte, J. Appl. Phys. 32:1536 (1961).
18. A. Taylor and H. Sinclair, Proc. Phys. Soc. (London) 57:108 (1945).
19. J. B. Nelson and D. P. Riley, Proc. Phys. Soc. (London) 57:160 (1945).
20. H. M. Otte and D. O. Welch, Phil. Mag. 9:299 (1964).
21. C. N. J. Wagner, A. S. Tetelman, and H. M. Otte, J. Appl. Phys. 33:3080 (1962).
22. J. A. Rayne, Phys. Rev. 112:1125 (1958).
23. J. A. Rayne, Phys. Rev. 115:63 (1959).
24. R. P. I. Adler, Doctoral thesis, Yale University, 1964.

Part II
Applied Research

Effects of Process Variables on the Properties of Molybdenum-TZM Alloy Sheet

W. A. McNeish

Refractomet Division, Universal-Cyclops Steel Corporation
Bridgeville, Pennsylvania

In molybdenum-TZM alloy, carbon is shown to be not simply a deoxidant, but an alloy addition affecting strength and recrystallization properties. The efficiency of the carbon can be varied by processing practices. For uniformity of sheet properties, carbon in the 0.020—0.030% range is desirable.

The strength of sheet can be varied as the amount of total cold work is varied, but the recrystallization temperature is affected very little within the permissible cold-work range to produce good quality sheet. Rolling and annealing temperatures, likewise, have little effect on recrystallization temperature. Brittle transition and formability are sensitive to degree of cold work, rolling temperatures, in-process and final annealing temperatures, and surface condition of the sheet.

The production of reproducible, high-quality sheet requires an understanding of the effects and control of all processing variables. When all variables are under control, sheet will be produced having recrystallization characteristics similar to those shown in Fig. 14 and having bend transition temperatures of below -50°F with good formability.

I. INTRODUCTION

Since the first commercial-sized ingot of molybdenum-TZM alloy was melted in 1959, continual efforts in process improvement and product evaluation have resulted in a corresponding improvement in sheet quality. Studies have been conducted on variables from ingot melting practice through each stage of processing to the final sheet annealing and surface conditioning. This paper summarizes some of the most significant process variables which contribute to final sheet properties and quality.

II. INGOT AND PRIMARY FABRICATION

Specifications for TZM alloy generally include the following chemical requirements: 0.40—0.55% titanium, 0.06—0.12% zirconium, and 0.010—0.040% carbon.

Titanium and zirconium are controllable within the specification range and can generally be held to about 0.50% titanium nominal ± 0.05 and 0.10% zirconium nominal ± 0.02. Unless both the titanium and zirconium are on the low end of the range, little variation in sheet properties is attributable to these alloy additions.

Carbon is of special interest. Variations within the 0.010–0.040% range will result in a significant variation in sheet properties. The solubility of carbon in molybdenum is approximately 0.018% at 4000°F and decreases rapidly to a few parts per million at room temperature. As a result, a network of grain boundary carbides forms in as-cast structures as shown in Fig. 1. This phase has been identified as primarily molybdenum carbide by Chang [1]. Retention of this phase can be traced throughout processing if improperly handled. This is illustrated in Fig. 2, which shows stringers of retained ingot carbides in $3/4$-in. intermediate material following extrusion, two rolling operations, and two recrystallization anneals.

The carbides cannot be arbitrarily eliminated by elimination of carbon additions or reduction to the minimum level required for deoxidation of the ingot (usually considered to be 0.010%). Figure 3 represents sheet produced from three ingots which were similar except for carbon content. Extruded billets from each of the three ingots were processed to sheet similar in all respects except for initial breakdown. One set of billets was hot–cold rolled to sheet bar from a 2200°F furnace; the second set was hot-forged to sheet bar above 3200°F. The effect of both carbon content and hot forging is evident in Fig. 3. Response of hardness to annealing temperature is an indirect measure of both strength and recrystallization temperature. The results indicate a precipitation hardening effect from carbon. The efficiency is increased by hot working. The effect diminishes with increasing carbon above about 0.020%.

The precipitation mechanism is apparent from standard free energy of carbide-formation curves (Fig. 4). The intersection of the titanium and zirconium free energy curves with molybdenum indicate that titanium carbide and zirconium carbide become more stable than molybdenum carbide below about 3150°F. As a result, working and annealing below this temperature does result in a very finely dispersed titanium/zirconium carbide phase. Examination of the hot-forged sheet bar structure (Fig. 5) indicates that the molybdenum carbide is refined and dispersed by hot working, thus improving the homogeneity and efficiency of the titanium/zirconium carbide phase developed in subsequent processing.

Figure 6 illustrates the broad range of sheet bar properties

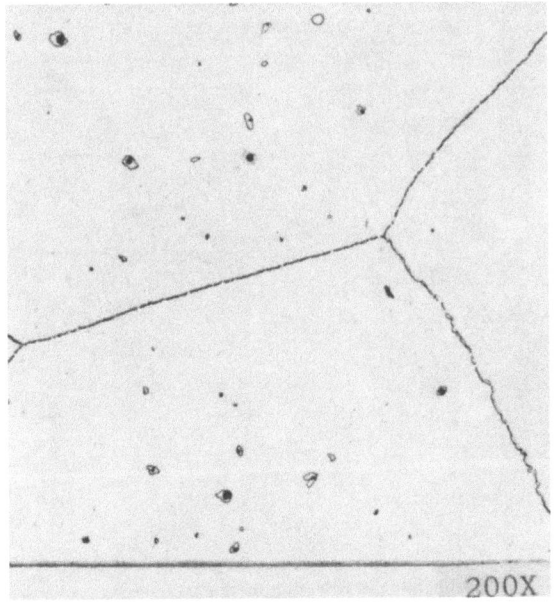

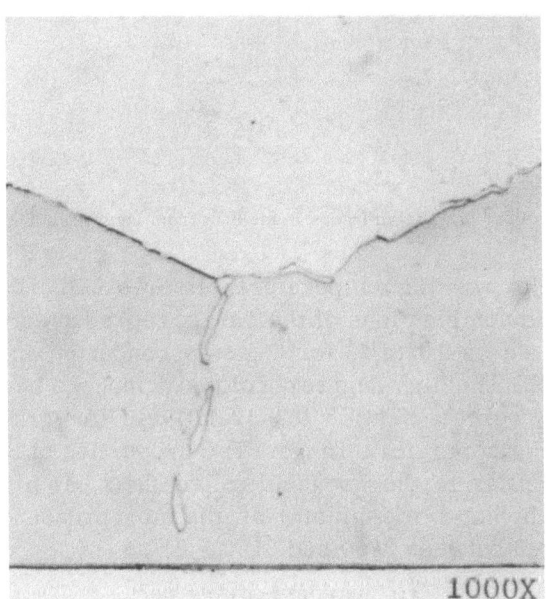

Fig. 1. As-cast TZM microstructure.

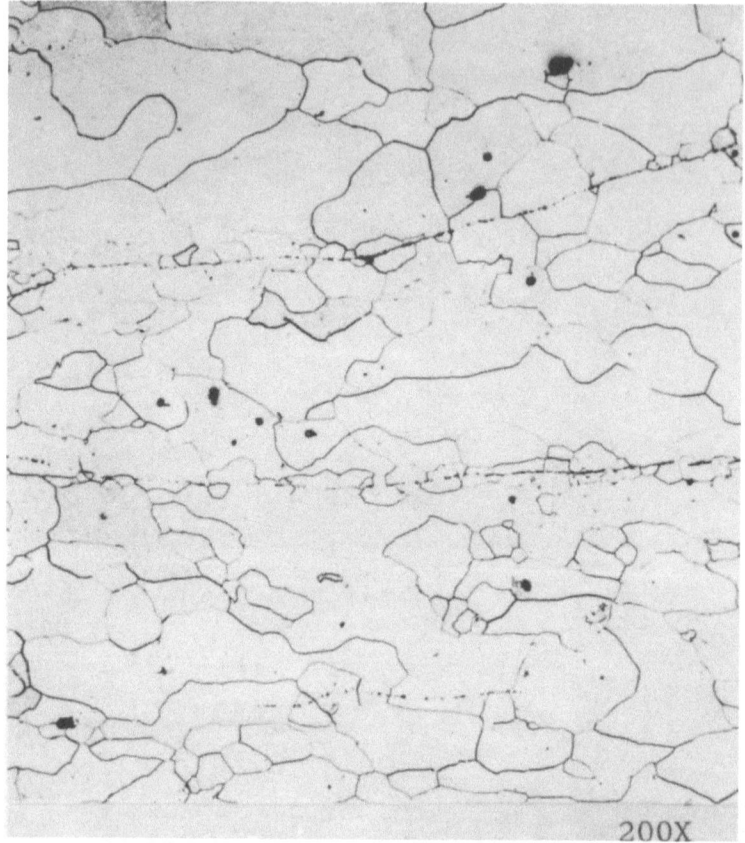

Fig. 2. Retained carbide stringers in moldout from ingot grain boundaries.

attributable to working temperature. Hot-worked and cold-worked sheet bar and combinations of the two are represented. In the sheet bar designated as 1098B4, forging was conducted in the range of 2320–1950°F. The high degree of cold work induced by this practice resulted in recrystallization to a relatively fine grain size in the 2600–2700°F temperature range. The properties of this sheet bar would be similar to the hot–cold rolled sheet bar previously discussed. In the hot-worked material, the most pronounced hardness drop occurred between 3000 and 3100°F.

III. INITIAL ROLLING

Reduction of sheet bar to finish sheet gage is accomplished by a

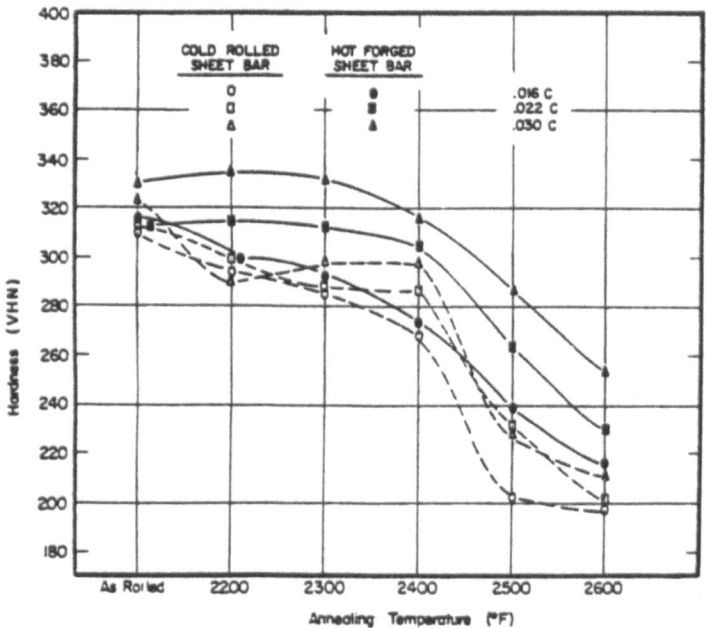

Fig. 3. Effect of carbon on recrystallization and hardness of 0.060-in. TZM sheet.

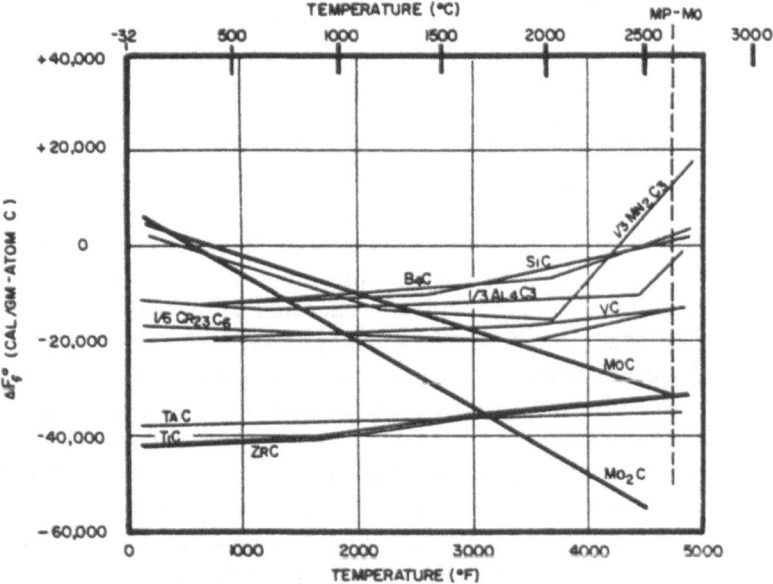

Fig. 4. Standard free energy of carbide formation for several elements soluble in molybdenum.

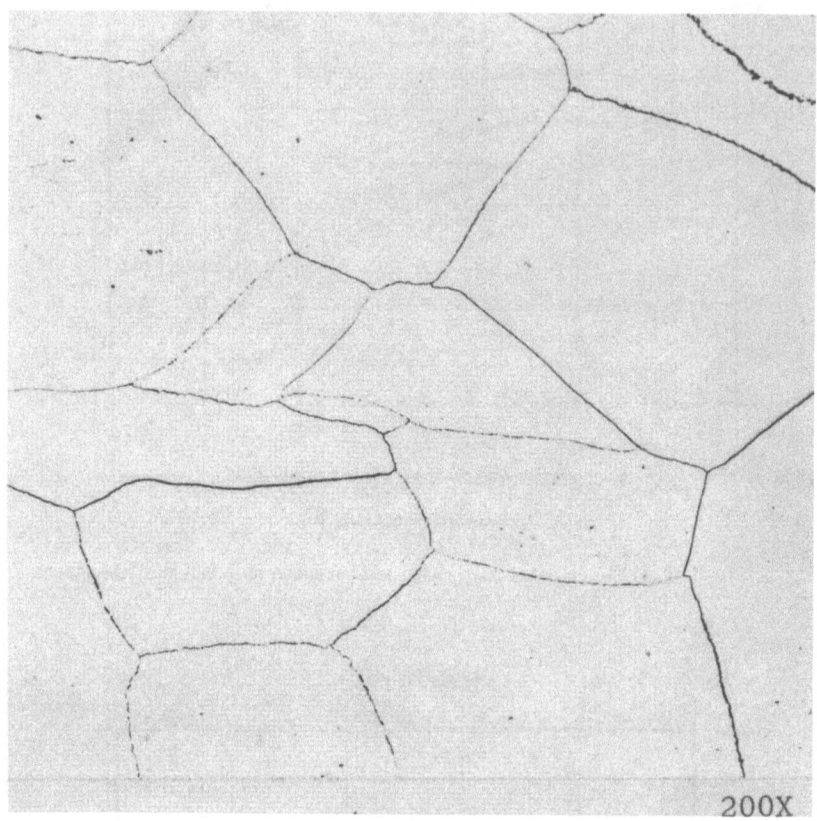

Fig. 5. TZM hot-forged sheet bar (3810—3200°F).

series of rolling and annealing operations designed to produce a
controlled degree of cold work which will result in optimum
properties. Initial rolling involves reduction of the sheet bar to an
intermediate gage which varies with the desired finish sheet gage.
A prime objective at this point is to produce a structure which will
recrystallize to a fine grain condition. This can be accomplished
by rolling in the 2000—2400°F range. Figure 7 presents recrystallization characteristic of material rolled to 0.125 in. from a 2000°F
furnace. Regardless of previous history of the sheet bar, recrystallization occurred between 2400 and 2600°F. Higher recrystallization
temperatures can be maintained by utilizing higher rolling temperatures. The results of rolling from a 2800°F furnace are shown in
Fig. 8. The recrystallization temperature was held approximately

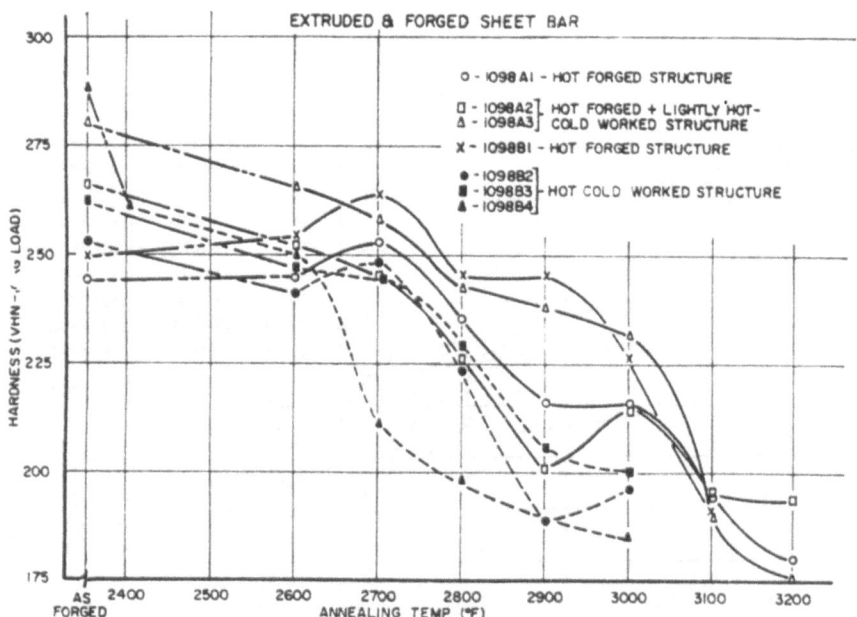

Fig. 6. Effect of annealing temperature on the hardness of extruded and forged sheet bar.

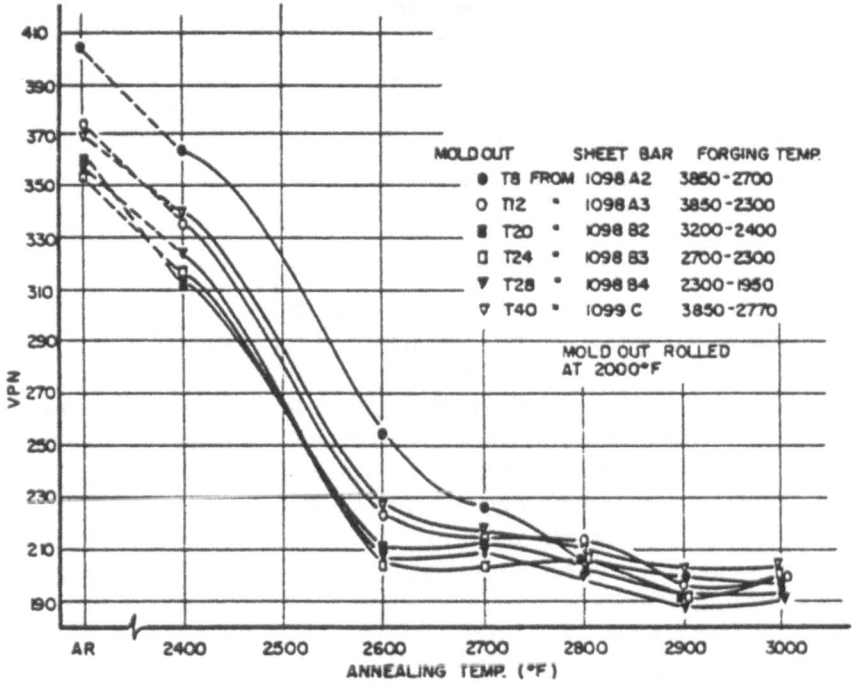

Fig. 7. Effect of annealing temperature on the hardness of nominal 0.125-in. moldout rolled at 2000°F.

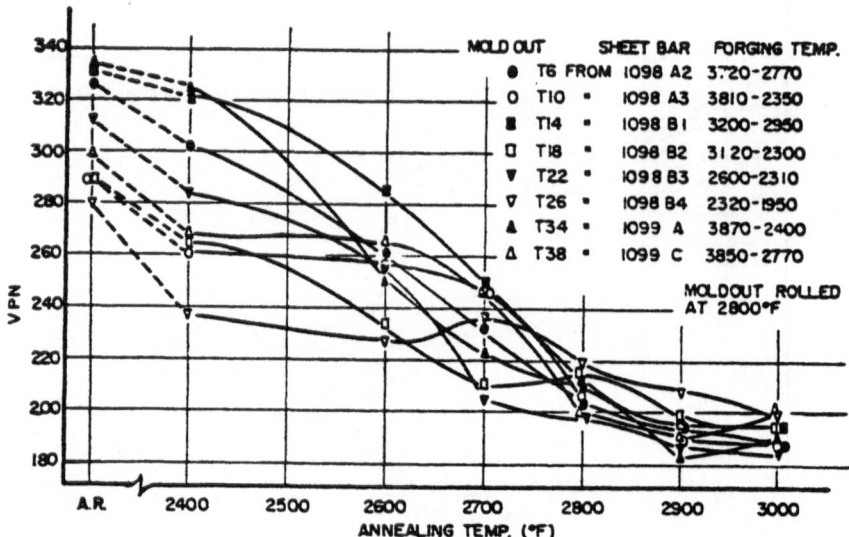

Fig. 8. Effect of annealing temperature on the hardness of nominal 0.125-in. moldout rolled at 2800°F.

200°F higher; however, a considerably coarser grain size resulted which adversely affected the brittle transition temperature of finished sheet. Note that the sheet bar forging variables influence the properties of the 2800°F rolled material.

IV. SHEET ROLLING

Figure 9 represents the recrystallization curve for 0.750-in. moldout rolled from $1\frac{1}{2}$ in. hot-forged sheet bar at 2200°F. Note that with this degree of cold work, initial recrystallization occurred at about 2450°F with full recrystallization at 2800°F. Material representing each of the points on this curve was rolled to 0.060-in. sheet. The resultant characteristics are shown in Fig. 10. The retained cold work in material annealed under 2800°F simply added to the total cold work resulting in the various hardness levels shown. Although a hardness variation of 50 DPH existed, traces of recrystallization were noted in all four materials following a 1-hr treatment at 2400°F.

A second series of sheet bars was rolled to various gages which would result in final cold work at 0.060-in. sheet ranging from 80 to 96%. Hardness levels varied in much the same manner as in the

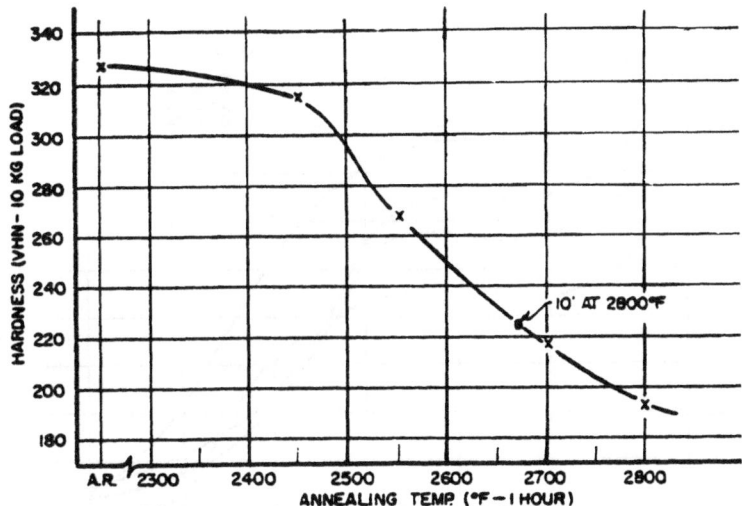

Fig. 9. Hardness of annealed 0.750-in. moldout

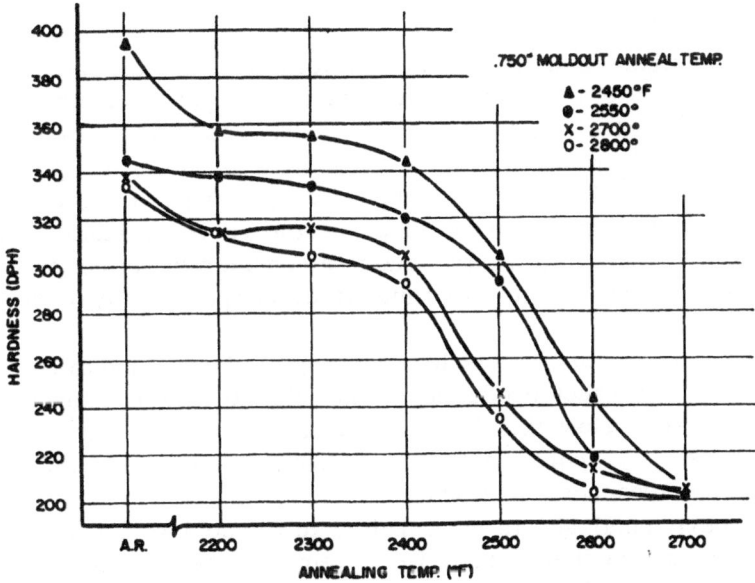

Fig. 10. Effect of 0.750-in. moldout annealing temperature on hardness of 0.060-in. TZM sheet.

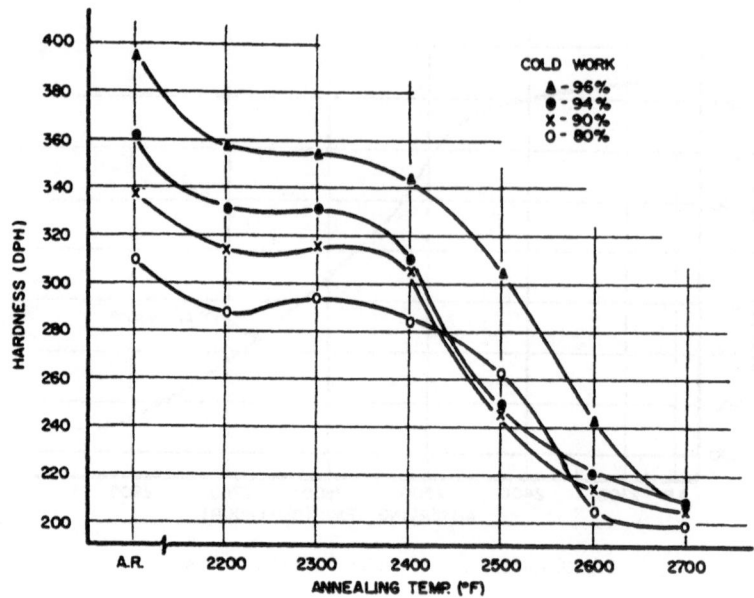

Fig. 11. Effect of percent cold work on 0.060-in. TZM sheet.

previous figure and, again, initial traces of recrystallization were noted at 2400°F in all cases. The sheet strength level can thus be varied by controlling the total cold work. The total cold work can be controlled by varying the intermediate annealing temperature (Fig. 10) or fully recrystallizing at varying gages (Fig. 11). The second method is preferred since partial recrystallization treatments are difficult to control due to a considerable effect of small deviations in temperature or time in the steep part of the recrystallization curve.

Total cold work at the level of 90% has been found to be most desirable for sheet production. As cold work is reduced to any great degree below this level, the brittle transition temperature increases. As cold work is increased above the 90% level, the lamination tendency increases.

Experimental sheets were rolled to 90% cold work with variations in intermediate stress relief temperature and final rolling temperature studied. Approximately half of the reduction was accomplished by rolling recrystallized moldout from a 2200°F furnace. Prior to conditioning and final rolling, stress relief anneals of 2150, 2300, and 2450°F were included. Final rolling from furnace temperatures of 1400, 1600, and 1800°F was then conducted on material in each of the stress-relieved conditions. Figure 12 shows that the intermediate stress-relief temperature had no effect on hardness or recrystallization of the finished sheet. In a similar manner, the final rolling temperature had essentially no effect on stress-relieved sheet hardness or recrystallization temperature (Fig. 13). An appreciable effect on bend properties of the sheet can be seen in Table I. All bends were conducted transverse to the final rolling direction. On an average, the 4T bend transition temperature varied directly with the moldout annealing temperature and the final rolling temperature. The minimum bend radius varied directly with the moldout annealing temperature; however, this property varied inversely with the final rolling temperature. The same tendency was found to exist when bending small panels over a less than 1T-radius at room temperature.

Surface layers of oxygen and/or nitrogen contamination will occur on TZM when processed at elevated temperatures in air. The contamination will retard recrystallization; thus, its presence and depth can be measured by flash annealing samples at 2600–

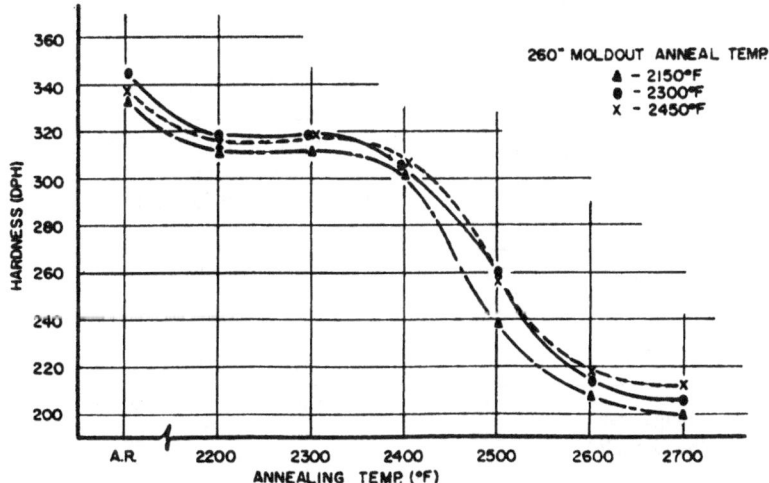

Fig. 12. Effect of 0.260-in. moldout annealing temperature on hardness of 0.060-in. TZM sheet.

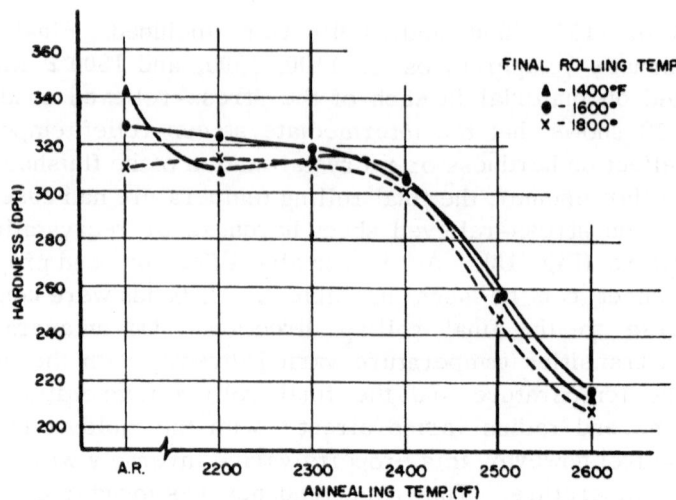

Fig. 13. Effect of final rolling temperature on hardness of 0.060-in. TZM sheet.

TABLE I

Effect of In-process Anneal and Final Rolling Temperature on Bend Properties of 0.060-in. TZM Sheet

	Moldout anneal	1400°F	1600°F	1800°F	Average
Bend transition (4T)	2150°F	−50°F	−50°F	−50°F	−50°F
	2300°F	−25°F	+25°F	0°F	0°F
	2450°F	0°F	+25°F	+25°F	+17°F
Average		−25°F	0°F	− 8°F	
Minimum bend radius (RT)	2150°F	0T	1T	0T	$^1/_3$T
	2300°F	2T	0T	0T	$^2/_3$T
	2450°F	2T	0T	0T	$^2/_3$T
Average		1 $^1/_3$T	$^1/_3$T	0T	
Flange test (bend angle)	2150°	75°	105°	120°	100°
	2300°	95°	80°	100°	92°
	2450°	87°	90°	92°	90°
Average		86°	92°	104°	

Note: All samples cold rolled 90% and annealed for one hour at 2300°F.

2700° F.   When examined metallographically, the unrecrystallized surface layer is evident.   This condition may run as deep as 0.003 in.   The above bend testing was conducted on material pickled free of surface contamination using HNO_3–HF solutions.   Recent work has indicated that, in inert atmospheres, impurities must be held in the low parts per million range to prevent surface contamination, which is measurable by metallographic and bend test techniques.

V. CONCLUSIONS

It has been shown that in TZM alloy, carbon is not simply a deoxidant, but an alloy addition affecting strength and recrystallization properties.   The efficiency of the carbon can be varied by processing practices.   For uniformity of sheet properties, carbon in the 0.020–0.030% range is desirable.

The strength of sheet can be varied as the amount of total cold work is varied, but the recrystallization temperature is affected very little within the permissible cold work range to produce good-quality sheet.   Rolling and annealing temperatures, likewise, have little effect on recrystallization temperature.   Brittle transition and formability are sensitive to degree of cold work, rolling

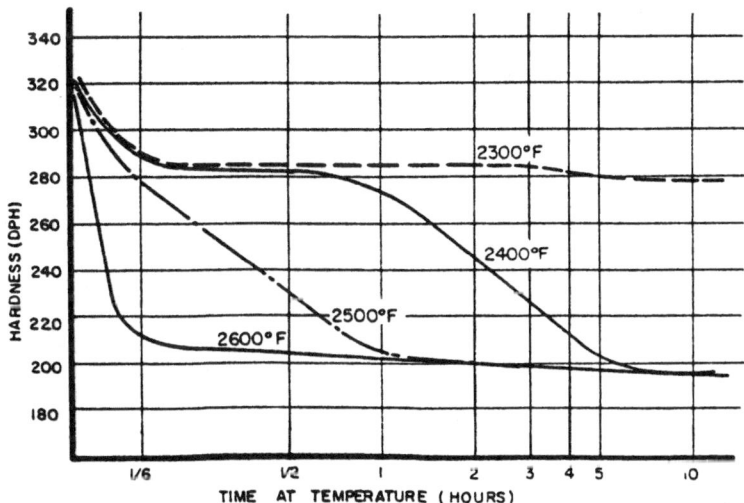

Fig. 14. Effect of time and temperature on recrystallization of TZM sheet.

temperatures, in-process and final annealing temperatures, and surface condition of the sheet.

It is apparent that the production of reproducible, high-quality sheet requires an understanding of the effects and control of all processing variables. When all variables are under control, sheet will be produced having recrystallization characteristics similar to those shown in Fig. 14 and having bend transition temperatures of -50°F or below with good formability.

ACKNOWLEDGMENTS

Portions of work included in this paper were conducted under Air Force contract AF 33(657)-8495 and Bureau of Naval Weapons contract NOas 59-6142-c.

REFERENCE

1. W. H. Chang, "A Study of the Influence of Heat Treatment on Microstructure and Properties of Refractory Alloys," Report ASD-TDR-62-211, Contract AF 33(616)-7125.

Arc-Cast Tungsten Sheet Plate and Bar

W. J. Schoenfeld

Universal-Cyclops Steel Corporation, Bridgeville, Pennsylvania

The production of tungsten base alloys has advanced rapidly in the past few years. In a comparison of powder metallurgy and arc-cast products, the potential as well as the observed benefits by the arc-cast process must be analyzed since neither of the two products is considered optimum. The purification and homogenization which occurs during melting should provide improved ductility and reliability in the end-products. In the preliminary work on W-25Re, it has been shown that alloying can greatly improve the ductile-brittle transition temperature (DBTT), particularly when the alloy is arc-melted.

I. INTRODUCTION

Tungsten has received considerable attention in space and missile applications because of its high-temperature strength potential. In these systems reliability is an essential, and the conventional powder metallurgy product is often questioned in this respect. Arc-cast tungsten products have a higher degree of reliability due to the purification and homogenization which results during the melting phase.

II. VACUUM CONSUMABLE ARC-MELTING

Arc-melting has two major advantages over the powder metallurgy approach in producing unalloyed tungsten products. First, the high-temperature arc operating in a dynamic vacuum results in volatization of impurity constituents the extent of which is controlled by the following:

1. The rate of gas throughput which is determined by the vacuum pump capacity and the cross-sectional area and the length of the path to the pumping system.
2. The melting rate which controls the time a particle is exposed to the vacuum while molten.
3. The vapor pressures of the impurity elements. At the arc temperature in tungsten melting, elements such as magnesium, sodium, potassium, and calcium having relatively

149

TABLE I

Percent Purification During Melting

	Mg	Na	K	Ca	Fe	Ni	Co
	(Results in parts per million)						
Electrode analyses	20	10	13	12	19	8	38
Ingot analyses	<1	<1	<1	<1	6	<1	<5
% Purification	>95	>95	>92.5	>91.8	68.5	>87.5	>87

The symbol < indicates analysis is below detection limits, and the symbol > indicates purification is greater than that indicated since the elements are below detection limits.

high vapor pressures are almost completely removed. Elements with a somewhat lower vapor pressure such as iron, nickel, chromium, and cobalt are readily removed but to a lesser degree, while the elements tantalum and molybdenum with vapor pressures approaching that of tungsten show very little change. This is more clearly illustrated in Table I where the percent purification for various elements is shown.

From Table I it could be assumed that regardless of the amount of magnesium, sodium, potassium, and calcium in the electrode, these elements would be completely volatized during the melting operation. Nickel and cobalt are also shown to be decreased below detection limits.

The second advantage is the larger size capability by the arc-cast route. For powder metallurgy sheet bar, the maximum starting size is approximately 100 lb, the limiting factor being the inability to maintain a uniform density as the cross-section increases. In PM wrought bar product, the size is currently very limited—the maximum diameter being approximately 1 in. and this having only a limited degree of work below the recrystallization temperature.

The typical arc-cast conditioned ingot shown in Fig. 1 weighed 640 lb. This 8-in.-diameter ingot can either be extruded to a nominal 4-in. diameter or a nominal 2 by 6 in. sheet bar.

A major advantage of arc-melting which will become more important in future tungsten work is alloy distribution. Homogeneous alloying in powder metallurgy is very difficult since the process is dependent upon solid state diffusion of the alloying elements during the sintering operation. In melting the alloy becomes relatively uniform during the molten state.

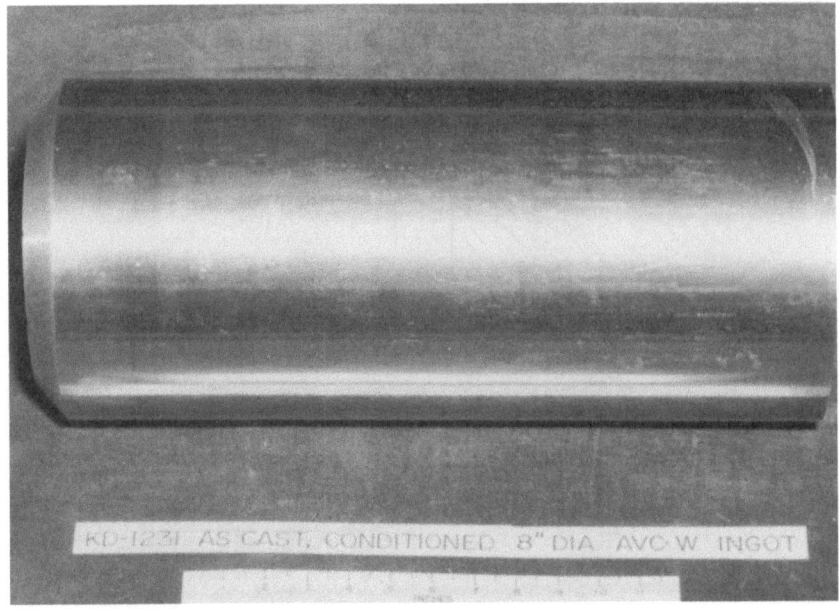

Fig. 1. Typical 8–in. diameter conditioned tungsten ingot.

III. PRIMARY BREAKDOWN BY EXTRUSION

The columnar grain structure resulting from the arc-melting operation presents a problem in initial deformation not found in the PM process. The interstitial impurity elements for which tungsten has a low solubility segregate to the grain boundaries. With the large grain size, the grain boundaries have a heavy concentration of impurities. These heavy grain boundaries readily fracture during conventional forging or rolling operations due to the tensile forces produced in these operations. The utilization of extrusion as the initial breakdown method has worked out very satisfactorily, since the billet is not subjected to tensile forces during the actual reduction period. This is illustrated in Fig. 2 which shows a schematic of the extrusion process as compared to forging. In the forging operation, the compressive forces on the billet ends result in secondary tensile forces on the perimeter. Attempting to forge tungsten in this manner would result in longitudinal surface cracks in the grain boundaries and/or opening up the center.

In the extrusion operation, the ingot is under hydrostatic compression except for the die opening. As the ingot is extruded, slight

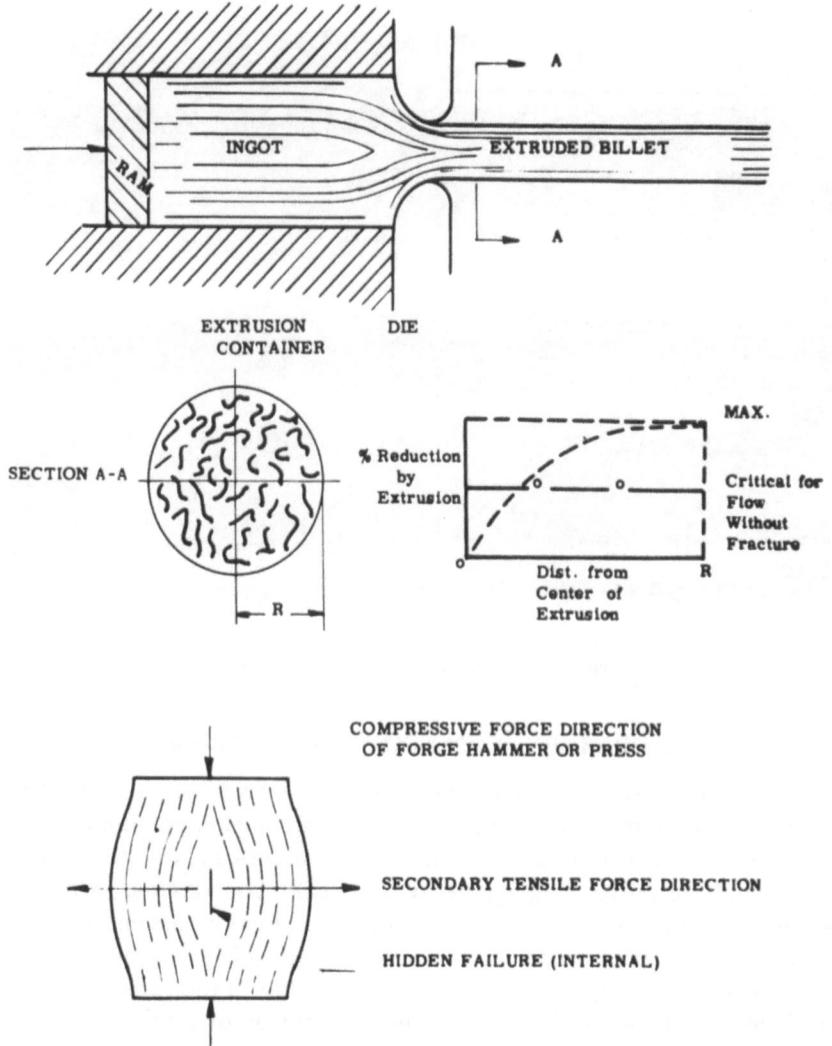

Fig. 2. Forces exerted in forging vs. extrusion.

tensile forces occur on the exit side of the die, but by this time the as-cast grain structure has been deformed providing a much larger grain interfacial area, thus minimizing the specific concentration of segregated interstitial impurities. The state of stress applied to material by the extrusion process, therefore, makes it the most favorable method for initial breakdown of low ductility materials such as tungsten.

IV. PRIMARY ROLLING FLAT PRODUCT

In the production of tungsten plate or as an intermediate rolling operation in producing sheet, it has been found that for both the arc-cast and PM product the rolling temperature should be the minimum which will permit reduction without fracture, since at excessive rolling temperatures grain growth occurs on reheating, which results in poor final properties, particularly strength and transition temperature.

The arc-cast material which has already received nominally 75% hot–cold work by extrusion can be rolled initially at 2300°F. The PM material requires temperatures in excess of 2700°F for initial rolling, so that, when starting with approximately a 1-in. thickness, at a $\frac{1}{2}$ in. thickness the piece has received little or no hot–cold work and, therefore, has an equiaxed grain structure which is still not full density. The tensile strength of this material at 900°F test temperature would be in the 50 – 60 ksi range, as compared to 80 – 90 ksi for the arc-cast material. Methods could be derived by which PM material would have a higher and possibly comparable strength; however, this is not the case under current technology.

V. SHEET ROLLING

The production of tungsten sheet requires consideration of the following characteristics: high-temperature strength, minimum ductile–brittle transition temperature, surface quality, flatness, gauge uniformity, and reproducibility.

TABLE II

Gauge and Flatness Evaluation

Gauge	Gauge variation		% Flatness	
	Target	Actual	Target	Actual
060	±0030	+0025 -0018	4	3.75
040	±0020	+0020 -0007	4	3.75
020	±0015	+0025 -0015	4	3.50

TABLE III

Tensile Property Evaluation

Gauge	UTS Variation (%)		0.2% YS Variation (%)	
	Target	Actual	Target	Actual
900°F (Longitudinal and Transverse):				
060	± 7	± 5.03	± 10	± 9.72
020	± 7	± 7.35	± 10	± 9.50
2000°F (Transverse):				
060	± 10	± 6.47	± 15	± 7.85
020	± 10	± 5.21	± 15	± 11.42

In unalloyed tungsten, strengthening and improved ductility are dependent upon strain hardening, since dispersion or solution strengthening mechanisms are not present. Within certain limitations, as the degree of cold work increases, the strength increases and the DBTT decreases. Under the conditions by which we roll sheet, the maximum reduction from the last recrystallization anneal is approximately 92%, since above that point delamination occurs and the properties deteriorate severely.

In the production of 36 by 36 in. sheets of 060, 040, and 020-in. gauge under an Air Force Contract,[*] the Materials Advisory Board (MAB) target properties were used as a criteria for evaluation of gauge control, flatness, and reproducibility of properties. Table II summarizes the results of the gauge and flatness evaluation.

It is shown that in all gauges except the 020-in. the actual gauge variation was less than the target value.

The 0.020 in. was shown to be out of tolerance a maximum of 0.001 in. In all cases, the flatness was less than the target maximum.

In Table III the variations in tensile properties are compared with the MAB target values. As shown the 0.020 in., 900°F UTS is the only one which exceeds the target values (7.00% as compared to 7.35%).

Tensile properties have been investigated on all three gauges from room temperature to 3000°F. The 0.060-in. gauge has received the most extensive testing to date, and the typical properties

[*]Tungsten Sheet Rolling Program—Contract AF 33(600)-41917.

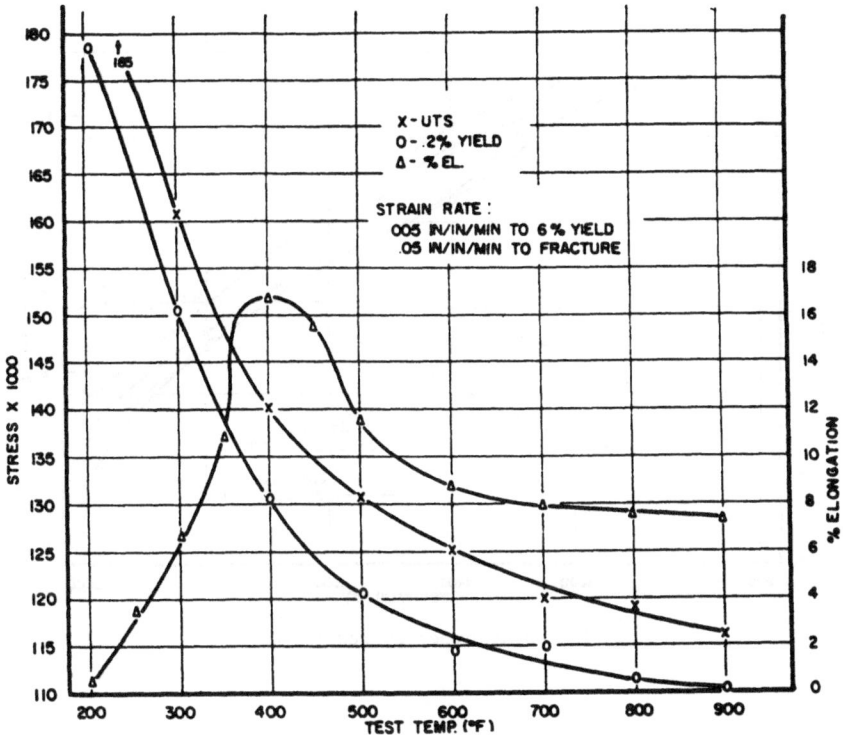

Fig. 3a. Low-temperature tensile properties (200–900°F).

of this material are shown in Figs. 3a, 3b, 3c, and 3d. As shown, at 200°F tensile ductility initiates with a corresponding UTS of 185,000 psi and 0.2% YS of 178,000 psi. The elongation which shows a peak of 17% at 400°F and then gradually decreases to 7% at 900°F, correlates with previously established data on powder metallurgy sheet. The UTS through this temperature range shows a gradual decrease to 117,000 psi. In the range from 900-2000°F, the UTS gradually decreases from 120,000 psi to a nominal 80,000 psi with a corresponding increase in elongation from 7 to 9%. Over the next temperature range, from 2000–3000°F, a severe drop in strength occurs between 2200°F and 2400°F. This is explained in Fig. 3d where the hardness is correlated with elongation. The rapid decrease in hardness over the same range is the result of increasing degrees of recrystallization during hold time prior to testing the samples. As explained earlier, tungsten depends upon strain hardening for strength, and this is lost when the material recrystal-

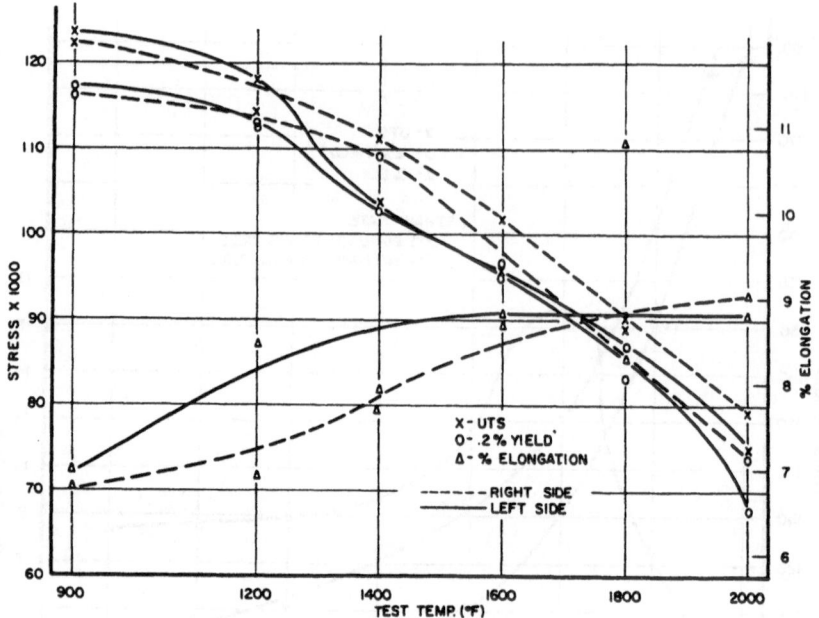

Fig. 3b. Tensile properties from 900–2000°F.

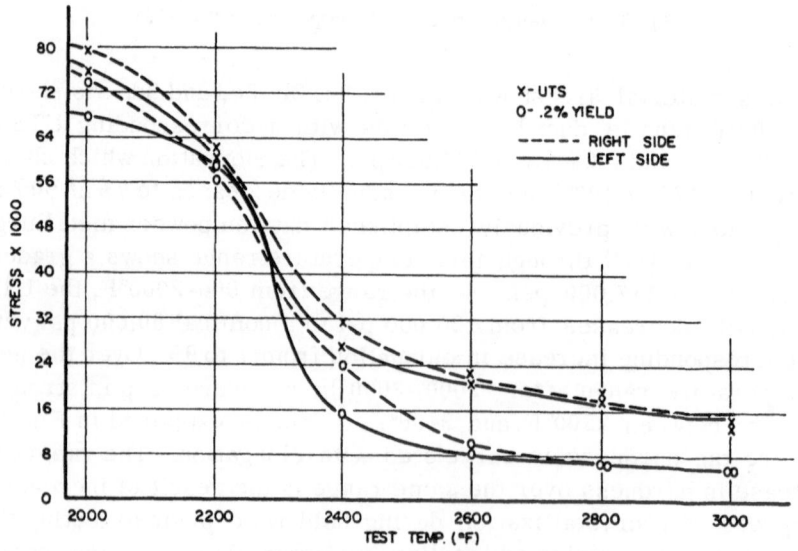

Fig. 3c. Tensile strength from 2000–3000°F.

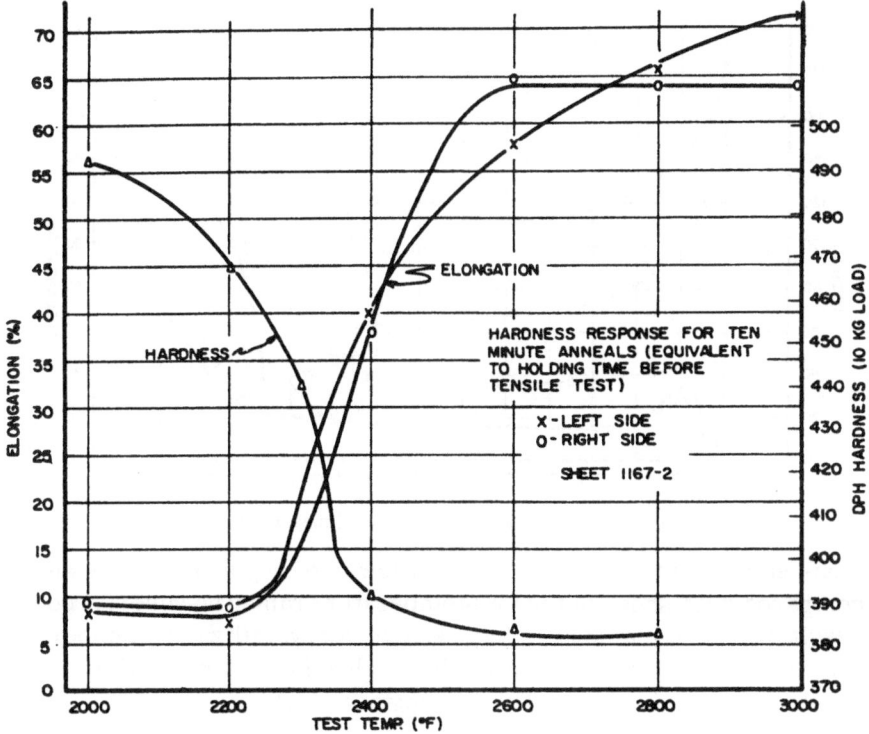

Fig. 3d. Tensile elongation from 2000–3000°F.

lizes. Conversely the elongation is shown to increase rapidly as recrystallization progresses. The tensile properties of 0.040 and 0.020-in. gauge are similar to those just discussed, with the exception that as the gauge decreases the low-temperature strength increases slightly with a corresponding slight decrease in elongation.

VI. BAR PRODUCT

Extruded rounds are used as the starting material for the production of bar. Most of the work to date has been with maximum extruded bar sizes of $1\frac{1}{2}$ in. diameter; however, they have received considerable reduction by extrusion, so that the starting bar is a wrought product.

The extruded product can be either rolled or swaged in the temperature range of 2300°F. In contrast to sheet rolling, it has been found that the working temperature cannot be lowered significantly during bar rolling without fracture of the material. This

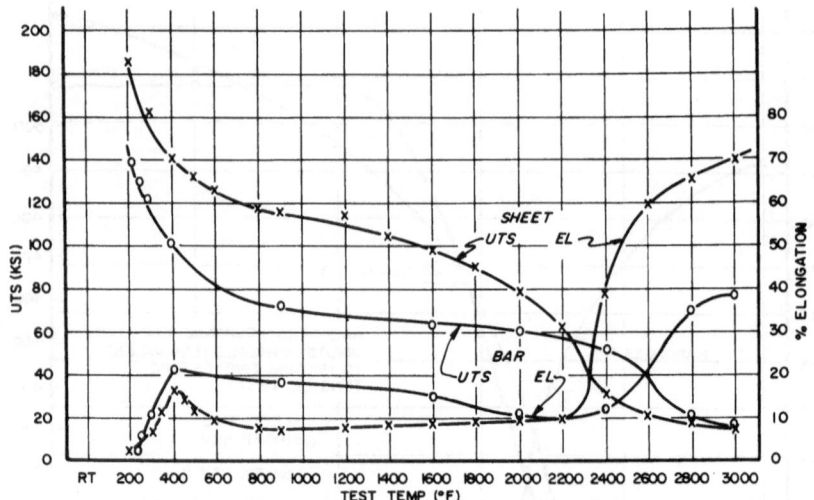

Fig. 4. Comparative strength and elongation of tungsten bar and sheet.

high-temperature working results in lower strength properties when compared with sheet in the low-temperature range; however, it does result in a nominal 300°F increase in recrystallization temperature for bar (nominal 2500°F). This is clearly illustrated in Fig. 4, which shows the comparative strength and elongation of arc-cast sheet and bar.

The most serious deficiency found in the powder metallurgy product to date is an extreme ductility loss above 3000°F. Figure 5

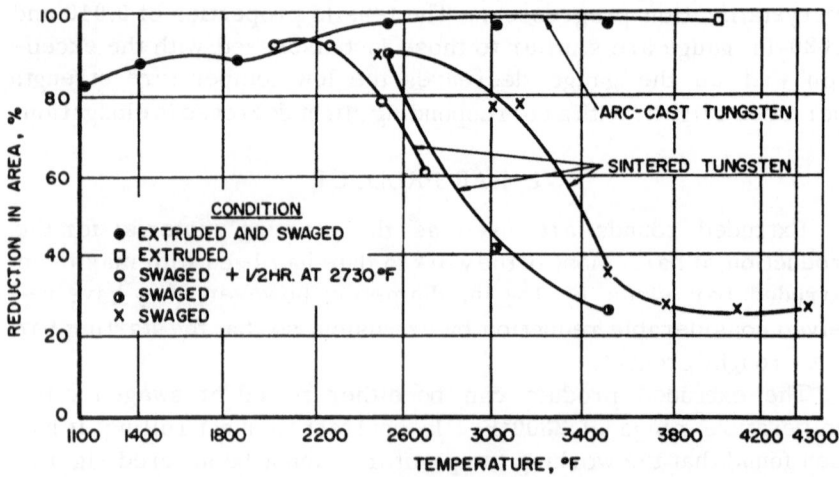

Fig. 5. Effects of temperature on the ductility of arc-cast and powder-metallurgy tungsten.

illustrates this effect. This work accomplished by NASA at the Lewis Research Center shows that arc-cast material maintains a reduction of area approaching 100% throughout the temperature range investigated, while the powder metallurgy product drops off severely to a low of approximately 25%. Considering the fact that tungsten is used primarily in the temperature range above 3000°F, this may present a serious problem.

VII. TUNGSTEN ALLOYS

Considerable work has been accomplished in development of tungsten alloys. Figure 6 shows the ultimate strength of selected alloys and unalloyed tungsten over the range of 2000–4000°F. As shown the tungsten–rhenium alloy has double the strength of unalloyed tungsten at 2000°F. At 3000°F it is still twice as strong as unalloyed tungsten; however, W-0.6%Cb has a higher strength at this temperature. The experimental alloy (W-20Ta-12Cb) is shown to have an exceptionally high strength at 3500°F.

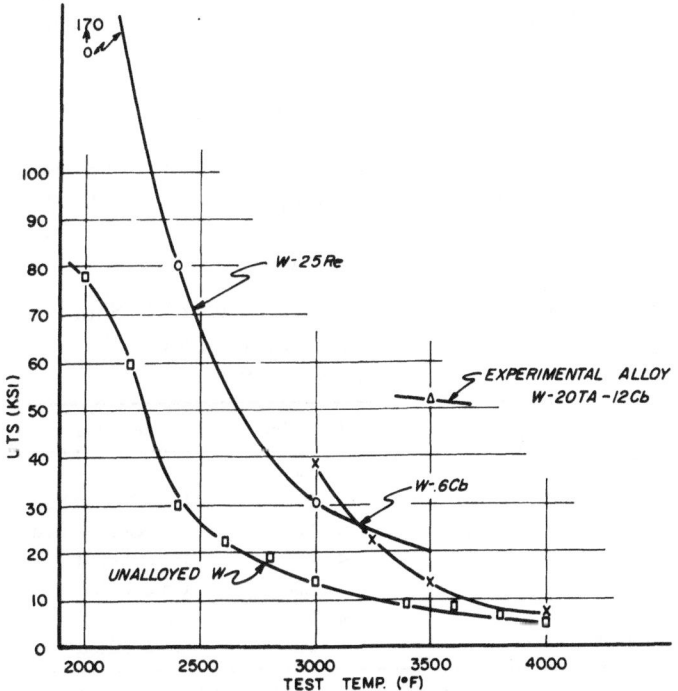

Fig. 6. Comparative UTS of unalloyed tungsten and selected alloys.

TABLE IV

Tensile Properties of Arc-Cast W-25 Re
(0.020-In. Gauge Cold Rolled Sheet)

Test temperature, °F	UTS·10³	0.2 %YS·10³	%EL
R.T.	236.1	229.8	7.4
900	165.3	163.4	2.0
2000	141.6	132.9	3.1
2400	100.9	92.6	17.4

The low-temperature properties of the tungsten—rhenium alloy are also outstanding. At room temperature, the reported yield strengths of wrought and recrystallized material are on the order of 400,000 and 300,000 psi, respectively. In preliminary investigations we have found the strength to be somewhat lower for this alloy, as shown in Table IV.

In sheet form the ductile—brittle transition has been observed as low as -300°F, as shown in Table V. Powder metallurgy W-25 Re in contrast is reported to have a DBTT range at ambient temperature in the wrought condition.

VIII. SUMMARY

The production of tungsten and tungsten base alloys has advanced rapidly in the past few years. In a comparison of powder metallurgy

TABLE V

Ductile—Brittle Transition of W-25 Re Sheet
(All Samples in Transverse Direction)
Bend Rate 1 in./min.

Gauge	T-Radius	DBTT (°F)	Remarks
0.125"	1	0	
0.030"	4	-225	Hot-Cold Rolled
0.020"	4	< -300	Hot-Cold Rolled
0.020"	4	< -300	Cold Rolled
0.020"	1	-225	Cold Rolled
0.015"	4	< -300	Cold Rolled
0.015"	1	-225	Cold Rolled

The symbol < indicates samples were not broken and brittle range was not determined.

and arc-cast product, the potential as well as the observed benefits by the arc-cast process must be analyzed since neither of the two products are considered optimum. The purification and homogenization which occurs during melting should provide improved ductility and reliability in the end-products.

In the preliminary work on W-25Re, it has been shown that alloying can greatly improve the DBTT, particularly when the alloy is arc-melted.

Ductile Chromium Alloys By Liquid-Phase Sintering

R. J. Van Thyne, R. E. Steiner, and F. C. Holtz*

IIT Research Institute, Chicago, Illinois

Ductility has been demonstrated in chromium-rich, two-phase alloys prepared by liquid-phase sintering. Microhardness was employed to follow work hardening of the chromium grains and the matrix. Matching of the flow stress of the two phases appears to be an important criterion. These principles have been successfully applied to other refractory metal systems.

I. INTRODUCTION

Chromium has been of interest for elevated temperature application for many years. Unfortunately, it exhibits a ductile-to-brittle behavior that generally results in room-temperature brittleness, similar to that of molybdenum and tungsten. The effects of alloying, grain size, prestrain, and extreme purity to produce ductile chromium under certain conditions are well-known. However, when melted or recrystallized to other than very fine grain size, the material—except for superpure chromium—is brittle at room temperature. Pickup of very small amounts of nitrogen during elevated temperature exposure to air readily embrittles chromium.

The objective of this program was the development of room-temperature ductility in chromium-base alloys by a structural approach. Two-phase alloys consisting of chromium grains surrounded by a thin alloy envelope of controlled composition were prepared by liquid-phase sintering. The objective of the alloying was a strong matrix so that the flow stresses of the two phases would nearly match; this is a simple but important deviation from past cermet work where a soft, ductile—but weak—matrix was generally the goal.

This initial study was to demonstrate the feasibility of the approach rather than to produce engineering alloys for a specific purpose. A matrix phase with a melting temperature lower than

*Present address: Central Scientific Company.

chromium may reduce the long-time properties, but the improved fabricability, ductility, and oxidation (nitrogen) resistance may more than offset a loss in mechanical properties. Further, it is not known that a loss in properties must result.

To produce the idealized microstructure, the phase equilibria must involve, at the sintering temperature, only slightly alloyed grains of chromium in equilibrium with a solute-base liquid that will provide a strong and ductile matrix when solidified. Alloy systems are selected to provide solubility of the chromium in the matrix at sintering temperatures. Hence, solution and precipitation occur during sintering and result in large grain growth and coherency between the matrix and grains. The ductile matrix surrounding each grain allows dissipation of dislocations during deformation so that slip can proceed. Other factors influencing the selection of alloy systems include relative flow stresses between the base and matrix materials, low solubility of alloying additions in the base material, and exclusion of compounds from the structure.

The binary systems of chromium with copper, gold, and silver generally provide the desired relationships although an intermediate phase occurs in the Cr–Au system. Work on the Cr–Ag materials was limited because of the high vapor pressure of silver at the sintering temperature. Initial studies were also performed on ternary additions to control the matrix properties of the Cr–Cu alloys.

II. EXPERIMENTAL PROCEDURES

Material Preparation

All powders utilized in this program were -325 mesh size. Preliminary studies using -150 mesh chromium powder proved difficult due to blending, pressing, and sintering problems. The primary constituent, calcium-reduced chromium powder obtained from the Lunex Company, was of 99.9% purity as were the copper and carbonyl nickel powders. The gold was 99.98% pure.

Material was prepared by powder metallurgical techniques. Powders blended by end-over-end tumbling in a glass jar were cold-pressed in either a $\frac{5}{8}$ in. diameter or $\frac{3}{4}$ by 2 by $\frac{1}{8}$ in. die. Pressing pressures in the range of 15–25 tons/in.^{2} were sufficient to yield good green strength in all compositions investigated. Raising or lowering the pressure failed to result in an improvement in sintered density.

Sintering was accomplished in a nonporous alundum tube in

a platinum-wound tube furnace. An atmosphere of pure hydrogen with a dew point of -70°F was provided by a gas train consisting of an oxygen catalyst, alumina drying tower, and liquid-nitrogen trap. The samples were packed in granular alumina in a molybdenum container. Temperature measurements were taken with an optical pyrometer through a sight glass at one end of the tube.

A number of sintering cycles were attempted; the most successful consisted of rapid heat to a temperature of 90°F below the liquid-phase melting point, a heating rate of 90°F/hr through the liquid-phase temperature, a rate of 180°F/hr to the sintering temperature, and a $\frac{1}{2}$ to 1 hr hold at temperature. All samples were furnace cooled. Sintering temperatures in the range of 2460°-2730°F were employed.

Suspected difficulties involving the reduction of chromium oxide in hydrogen failed to materialize. A fluxing alloy of Ag–1 w/o Li was tried in the binary Cr–Cu alloys; there was very little, if any, improvement in density. A substantial increase in both strength and ductility was achieved by employing prereduced copper powder.

Material Evaluation

Rectangular specimens $\frac{1}{8}$ by $\frac{3}{16}$ by $1\frac{3}{4}$ in. machined from the 2 by $\frac{3}{4}$ by $\frac{1}{8}$ in. sintered blanks were tested for room-temperature strength and ductility by means of a modified transverse-rupture test. The specimen was supported on parallel cylindrical pins having a span of $1\frac{1}{4}$ in. A load was centrally applied, and transverse-rupture strength values for the materials were calculated from the formula

$$s = 3PL/2wt^2$$

where s is the transverse rupture strength, psi; P, the applied load, lb; L, the length between supports, in.; w, the width of the sample, in.; t, the thickness of the sample, in.

Although the values obtained for transverse-rupture strengths provided only a clue to the ultimate tensile strength, the comparative values of this method were sufficient for screening purposes. Ductility, expressed in inches of deflection, also provided good comparative data. It should be noted, however, that variation in sample thickness will result in a variation of deflection. Hence, thinner samples will exhibit a greater degree of ductility due to the less severe bend.

Vickers microhardness data were taken with a 25-g load applied to a diamond indenter. An understanding of relative flow stresses of both matrix and chromium was obtained through hardness values. Also, hardening of the chromium by impurities or solute elements was readily determined.

III. RESULTS

The matrix volume was varied for each system and the best continuity and distribution of matrix occurred at about 20 vol. %. Most of the data are presented for this level, but all compositions are reported in weight percent.

Sintered densities of the materials investigated ranged from about 90 to 95% of theoretical. In general, as the matrix volume was increased, the density was improved. Since theoretical density was not obtained, the property data are not absolute but are helpful for comparative purposes. Fully dense materials could be developed with additional work and would be necessary to obtain optimum ductility and fabricability.

A cursory rolling study indicated relative ductility of the materials. Although edge cracking was severe with most compositions, some materials were rolled to over 90% reduction in thickness. The as-rolled materials had very low ductility. When the materials were rolled, small transverse cracks sometimes appeared during early stages of reduction and eventually propagated into the material. It is unlikely that any of the chromium materials could be produced in reasonably sized thin sheets at the present stage of development.

Chromium–Copper System

Binary compositions of Cr–Cu in the range of 10–35 w/o copper were studied. Figures 1 and 2 illustrate the microstructures of unalloyed chromium and a Cr–30 w/o Cu alloy as-sintered. Figure 2 shows the large chromium grains in a matrix of almost pure copper, since the room-temperature solubility of chromium in copper is only about 0.7 w/o. Note the large amount of grain growth as compared to the unalloyed chromium. This alloy was cold rolled to over 90% reduction in thickness. Figure 3 shows the extensive deformation of chromium grains although severe edge cracking occurred and the as-rolled material was brittle. Original ductility can be restored by resintering the worked material to its original microstructure.

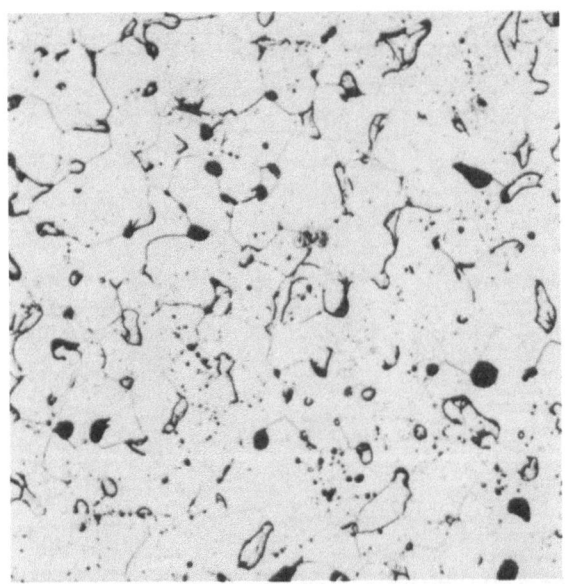

Fig. 1. Unalloyed chromium sintered in hyrdogen at 2550°F for 1 hr. Etched electrolytically with a 5% solution of aqua regia. ×250.

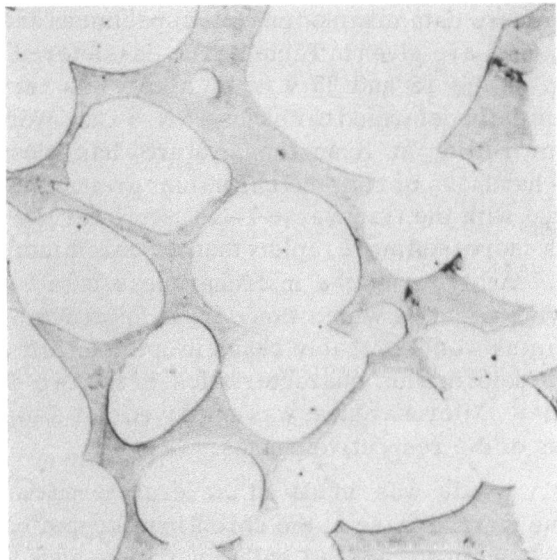

Fig. 2. Cr—30 w/o Cu sintered in hydrogen at 2550°F for 1 hr. Chromium grains in a copper-rich matrix. Note grain size as compared with Fig. 1. Unetched. ×250.

Fig. 3. The specimen shown in Fig. 2 cold-rolled 90%. Grains have been severely deformed while retaining continuous distribution of matrix. Unetched. ×250.

Microhardness data obtained on rolled specimens and transverse 1–1 rupture bars are given in Table I. The as-sintered hardness of both phases in the 15 and 25 w/o Cu alloys was very similar to that shown for the undeformed bar of Cr–25 w/o Cu. Working, either by bending or rolling at room temperature, has substantially increased the hardness of both the chromium grains and the matrix. It appears that with the transverse 1–1 rupture bar the hardness of the matrix is increased more rapidly than the chromium grains upon deformation. Analyses of the microhardness data indicated that material having a matrix with a flow stress approaching that of the chromium grains would probably result in optimum fabricability and ductility; the deformation characteristics of the two phases would then be similar. Microhardness was employed as a simple measure of flow stress of the respective phases.

A cursory study was made of several ternary additions to strengthen the matrix phase of the chromium–copper base. Molybdenum, nickel, manganese, and palladium were added to increase the solubility of chromium in the matrix. A matrix volume of 20 to 25% was maintained in the ternary alloys, and improved structures

TABLE I

Microhardness Data for Chromium–Copper Alloys

Composition, w/o	Condition	Microhardness, VPN	
		Cr grains	Matrix
Cr – 25Cu	Undeformed portion of bar	155	85
Cr – 25Cu	Fractured portion of bar	180	140
Cr – 15Cu	Rolled 50% at room temperature	195	125

were produced with each of the additions. Grain growth was greater, and the distribution of matrix was improved.

All of the ternary additions resulted in a decrease in strength and ductility as compared to the binary Cr–Cu alloys. Nickel in amounts up to 2 w/o resulted in a catastrophic loss of properties. The Cr–Cu–Ni alloys could not be rolled at room temperature, as severe cracks formed with initial reductions. Microhardness data given in Table II clearly explain the poor deformation characteristics. Although nickel additions were useful in increasing the low strength of the copper matrix, the embrittlement was a result of increased solubility in the chromium. Molybdenum and manganese added in the range of 0.25 to 5 w/o produced generally similar though less severe results. Palladium, despite the presence of the compound Pd_2Cr_3 in the binary Cr-Pd system, was added to the Cr–Cu system over the range 2.5 to 20 w/o of the total alloy content. The presence of copper suppressed the formation of the compound but did not eliminate it entirely, and the materials were brittle.

Chromium–Gold System

Chromium–gold binary alloys in the range of 12.5 to 50 w/o Au (5–25 v/o Au) were investigated. Sintered densities ranged from 90 to 95% of theoretical. The microstructure of a Cr–40 w/o Au (Cr–20 v/o Au) alloy is shown in Fig. 4. Transverse-rupture data for this material showed good strength and moderate ductility as compared to the previous alloy systems investigated. Rolling at room temperature with reductions to 60% in thickness was possible without noticeable edge cracking although bend ductility was low. The hardness of the two phases was more closely matched with the Cr–Au system than with the Cr–Cu materials as shown in

TABLE II

Mechanical Properties of Chromium Alloys

Composition, w/o	Microhardness, VPN		Transverse rupture		Deformation (room temperature)
	Cr grains	Matrix	Strength, ksi	Deflection,* in.	
Cr – 25Cu	150	80	62	0.120	Good
Cr – 18Cu – 2Ni	360	110	36.1	0.020	Poor
Cr – 40Au	170	140	90.4	0.080	Very good
Cr – 40Au†	165	190	75.5	0.100	--

*Deflection over a $1\frac{1}{4}$-in. span.
†Exposed to 1830°F air for 35 hr.

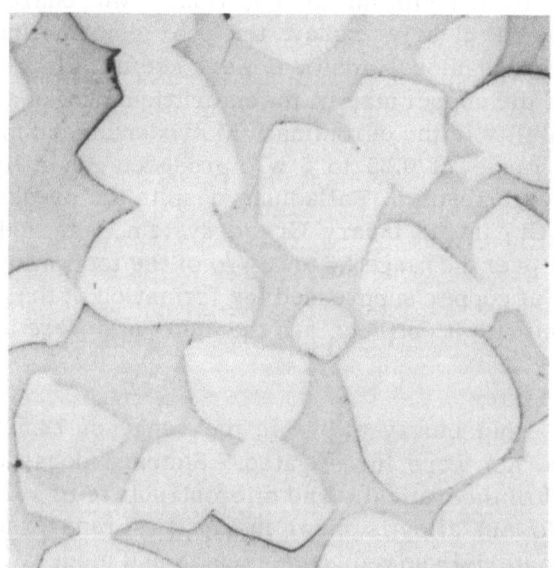

Fig. 4. Cr—40 w/o Au alloy sintered in hydrogen at 2550°F for 1 hr. Chromium grains in a matrix of Au—Cr. Unetched. ×250.

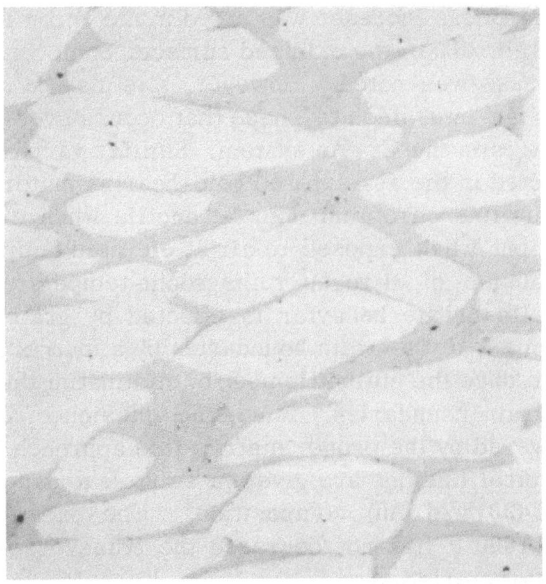

Fig. 5. The specimen shown in Fig. 4 cold-rolled 70%. Unetched. ×250.

Table II. Figure 5 illustrates the structure of the Cr–40 w/o Au alloy rolled to 70% reduction in area. Deformation of the chromium grains is readily apparent, but reduction in thickness was not carried to the same extent as the Cr–Cu materials. No direct comparison may be made between the deformation capability of the two binary systems. However, in both systems porosity must be eliminated to achieve maximum properties.

Oxidation Resistance of Chromium–Gold

The Cr–Au alloys were exposed to 1800°F air, and the weight change was determined after 2 and 4 hr. Unalloyed chromium exhibits a slight weight increase (0.04 mg/cm^2/hr) at this temperature [1] and the Cr–40 w/o Au alloy showed similar changes. However, the materials containing lower amounts of matrix phase exhibited higher weight increases. This may be due to diffusion of the contaminants in the matrix followed by more extensive reaction with the large surface area of the individual chromium grains.

Microhardness data were obtained on a number of specimens after exposure to air at 1700° to 1830°F for times up to 35 hr. Very little or no hardness increase was noted in the chromium grains at distances within 25 μ of the oxidized surface. Some variation in the matrix hardness was noted. However, this may be related to the formation of an intermediate phase that occurs over certain temperature ranges in the Cr–Au system. Similar variations in hardness were noted in the as-sintered gold-bearing materials.

Polycrystalline chromium can be ductile when of sufficiently high purity, but when exposed to air at elevated temperatures the resulting infusion of nitrogen ruins room-temperature ductility. This ductile-to-brittle behavior is affected by grain boundaries. The replacement of the grain boundaries by a matrix alloy may be expected to reduce the embrittlement by eliminating the chromium–chromium grain boundaries. Also, the infusion of contaminants could be lowered by the proper matrix; this approach has not been studied. Typical findings are given in Table II for an exposed Cr–40 w/o Au (20 v/o Au) composition. The oxidation (nitrogen) exposure at 1830°F has not degraded the transverse-rupture deformation characteristics.

Other Refractory Metal Systems

According to the principles developed during a series of investigations, good deformation was observed in the simple binary chromium alloys because the hardness of the chromium grains and the matrix was sufficiently similar. This hypothesis concerning the need for closely matching flow stress—as measured by microhardness—was critically tested during studies of other refractory metals having similar envelope structures.

The development of fabricable tungsten-base alloys serves to further illustrate these principles of ductilizing liquid-phase sintered materials. Structurally, these alloys consist of rounded tungsten grains in a solid-solution matrix similar to Fig. 2; matrix volume is about 20%. The commercial W–6 w/o Ni–4 w/o Cu alloys exhibit poor ductility (Table III). Although the matrix is quite soft, a serious hardness mismatch occurs between the two phases. The composition W–10 w/o Ni also possesses poor ductility because the hardness of the matrix is too high. However, with W–6 w/o Ni–4 w/o Fe, the hardnesses of the phases are more similar, and excellent ductility results. This composition exhibits 20 to 25% tensile elongation at room temperature. In all three of these examples, the microhardness of the tungsten grains is nearly equal. The microstructure of W–Ni–Fe rolled to a reduction in thickness of 90%

TABLE III

Mechanical Properties of Envelope-Structure Tungsten-Base Alloys

Composition, w/o	Microhardness, VPN		Transverse rupture	
	Tungsten grains	Matrix	Strength, ksi	Deflection,* in.
90W – 6Ni – 4Cu	440	260	235	0.05
90W – 10Ni	435	560	135	0.02
90W – 6Ni – 4Fe	430	350	260	0.6 †

*Transverse rupture deflection of $1\frac{1}{4}$-in. span for sample thickness of about $\frac{1}{8}$ in.
†Maximum deflection corresponding to about 180°, $1.5t$ bend. $1.5t$ is a method of describing bend-test data, where t refers to the thickness of the material.

shown in Fig. 6 corresponds to Fig. 3. A similar microhardness analysis has been successfully applied to liquid-phase sintered molybdenum alloys and a deformed structure is given in Fig. 7.

IV. SUMMARY AND CONCLUSIONS

Room-temperature ductility has been developed in chromium alloys by means of an envelope-type of microstructure produced by liquid-phase sintering techniques. Binary chromium-base alloys with copper and gold provided the required phase equilibria, and the materials generally contained about 20 v/o matrix. Ternary solutes were included to strengthen the copper matrix and improve ductility, but the additions were generally detrimental.

Metallographic analysis of materials rolled at room temperature to reductions in thickness of 90% showed extensive deformation of the chromium grains. Ductility was also shown for Cr–Cu and Cr–Au by transverse-rupture testing. The transverse-rupture ductility of Cr–Au was not reduced by exposure to 1830°F air for 35 hr. This investigation did not proceed to the stage where sizable sheet was produced. Sintered densities of 90 to 95% were generally achieved, and further processing development to obtain near-theoretical density would be necessary to optimize fabricability and ductility.

One of the significant results was the conclusion based on these and related studies that in many two-phase materials strengthening

Fig. 6. W—7 w/o Ni—3 w/o Fe alloy liquid-phase sintered
and cold-rolled 90%. Longitudinal section showing elongated
tungsten-rich grains. Unetched. ×250.

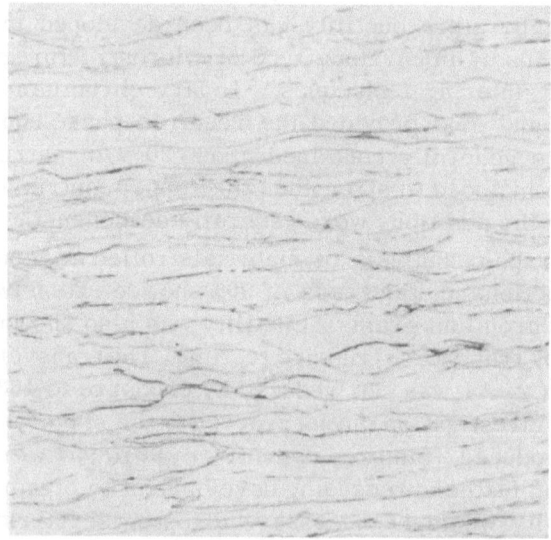

Fig. 7. Liquid-phase sintered molybdenum alloy cold-rolled
90%. Deformed molybdenum grains in a discontinuous
matrix. Unetched. ×500.

of the matrix phase will improve ductility and fabricability. Deformation of transverse-rupture bars and rolled specimens was followed by microhardness measurements of the chromium grains and the matrix. The principle of matching the flow stress of the two phases (indicated by microhardness) was demonstrated. These techniques were successfully applied to liquid-phase sintered tungsten and molybdenum compositions also containing about 20 v/o matrix. W–Ni–Fe alloys could be readily cold rolled and exhibited 20 to 25% tensile elongation at room temperature.

Ductile chromium alloys containing a modified matrix phase may find certain aerospace applications as structural materials or as coatings because of the improved ductility, fabricability, and resistance to degradation by exposure to elevated temperature air. More extensive evaluation of related two-phase tungsten alloys has shown that good elevated temperature tensile strength may be obtained with such structures; the long time strength is more seriously reduced. Whereas the 2000°F tensile strength of W–6 w/o Ni–4 w/o Fe (with a 20 v/o matrix melting at 2435°F) is 27,000 psi, the 100-hr rupture strength is about 2500 psi. Higher strengths are obtained by further alloying or by reducing the volume of the matrix. The alloys prepared for this initial investigation of chromium-base alloys had relatively low melting matrices, but higher melting ternary alloys containing noble metal additives could be developed. This should result in improved elevated temperature strength while retaining the advantages of the two-phase structure.

ACKNOWLEDGMENTS

The work on chromium alloys was sponsored by the IIT Research Institute, and the authors appreciate the opportunity to publish these findings. Many of the experimental data were obtained by Mr. James Ivans.

REFERENCE

1. C. A. Krier, "Coatings for the Protection of Refractory Metals From Oxidation," Defense Metals Information Center Report 162, November 24, 1961.

Formable Sandwich Structures for Aerospace Applications

H. R. Ogden and J. A. Houck
Battelle Memorial Institute, Columbus, Ohio

L. H. Abraham
Douglas Aircraft Co., Santa Monica, California

and R. I. Jaffee
Battelle Memorial Institute, Columbus, Ohio

A process has been developed for the production of sandwich structures which
can be formed to complex contours without detrimental effects to the bond
juncture between the core and face sheets. The sandwich structures are produced by roll-bonding, a process which results in a reliable diffusion bond
between the core and face sheets which has the properties of the base metal.
To support the sandwich structure during rolling, the voids in the sandwich are
filled with a material which can be preferentially removed after roll-bonding
and forming have been completed. For titanium alloys, which are well-suited
to this process, mild steel is used as the filler material. The steel is removed
as a last step in the process by preferential leaching in a dilute nitric acid
bath. After roll-bonding and with the steel filler still in place, samples of
Ti-6Al-4V truss core sandwich have been bent to a radius of 2T without failure
at the normal forming temperatures of 1400–1600°F for this alloy. Compound
curvatures, such as dome sections, have been formed from this same material.
A wide variety of core configurations can be bonded by this process. The
chief limitation to the core design is that there must be continuity of the filler
metal to permit its removal.

I. INTRODUCTION

The advantages of high strength and stiffness combined with light
weight to be gained by the utilization of sandwich structures in aircraft and space vehicles has long been recognized. The chief
deterrents to the use of sandwich structures have been the lack of
formability and the problems involved in the production of large
sandwich panels.

The roll-welding process described in this paper is capable of
producing large sandwich panels which can be formed into compound curvatures. The essential features of the process are the
diffusion bonding of the face sheets to the core during a hot rolling

177

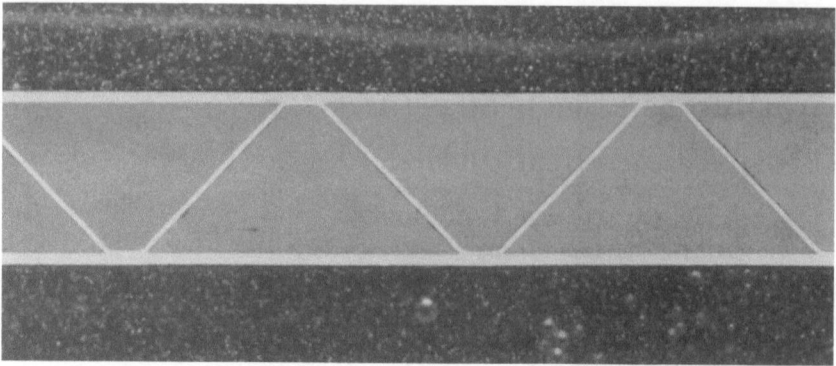

Fig. 1. Cross section of roll-welded Ti-6Al-4V truss-core sandwich (5 ×, reduced 30% for reproduction).

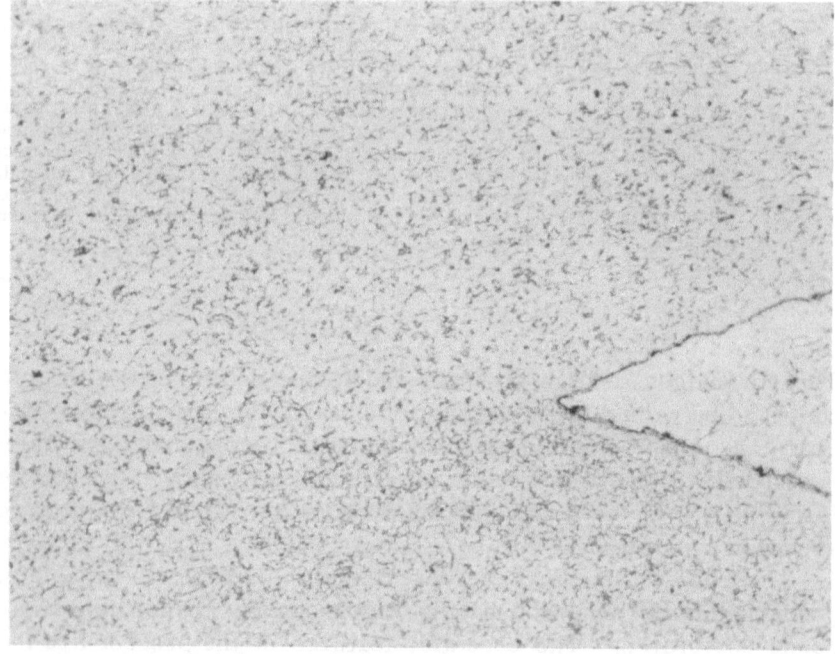

Fig. 2. Photomicrograph of bond area between core and face sheet of Ti-6Al-4V sandwich (500×).

operation, with complete support for the structure provided by a dissimilar metal container and matrix, forming operations being conducted with the matrix still in place and removal of the matrix material by a preferential leaching operation.

II. PRODUCTION OF SANDWICH STRUCTURES

The roll-welding process for the production of sandwich structures takes advantage of the interdiffusion of metals that occurs through the application of deformation, temperature, and time during a hot rolling operation. This is essentially the same type of diffusion bonding that takes place when roll-cladding one metal to another. In the manufacture of sandwich structures, however, the voids in the sandwich are filled with a dissimilar metal which can be removed later. Figure 1 is a photograph of a cross section of a Ti-6Al-4V truss-core sandwich with the mild steel matrix material still in place, while Fig. 2 is a photomicrograph of one of the bond areas between the core and the face sheet. It is seen that this process yields a uniform core configuration with a complete metallurgical bond between the core and the face sheets.

Since this is a deformation process in which the thickness of the sandwich panel is reduced during hot rolling, the sandwich design that is desired must be translated to a starting pack design taking into account the reduction to be accomplished during rolling and the rolling direction. For example, the 45° truss-core sandwich shown in Fig. 1 was designed to have the dimensions shown in Fig. 3a. The starting pack design, then, for a hot rolling reduction of 60% and rolling in the longitudinal direction of the corrugations is shown in Fig. 3b. Note that the horizontal members of the pack (the face sheets) are reduced 60% during rolling, while the angle and the thickness component of the truss-core are reduced in proportion to the angle of the truss.

Having established the starting pack design, the pack assembly is prepared. For example, an exploded view of a laboratory-size pack is shown in Fig. 4. A yoke is used to contain the assembly which is composed of the core, filler material, and cover sheets. The core may be corrugated as shown in Fig. 4 or layed up in individual strips depending upon the complexity of the core design. Pack cover plates are welded to the yoke to form an air-tight seal, and the pack is hot-evacuated prior to rolling.

Rolling is done on standard plate mills to the desired total

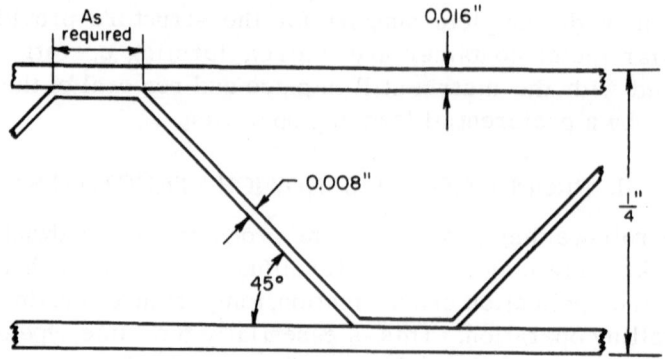

a. Desired configuration

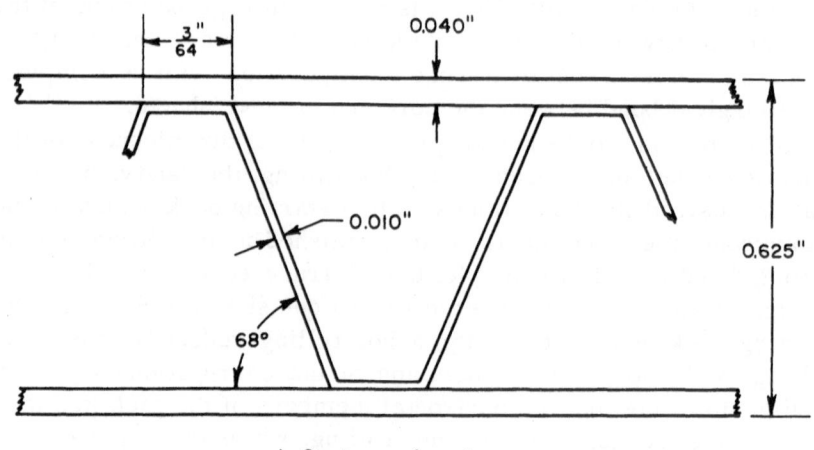

b. Starting configuration

Fig. 3. Design configurations for a truss-core sandwich panel.

reduction, at temperatures normally used for the hot rolling of the sandwich materials. It is necessary to provide one or two reheats in the rolling schedule to ensure sufficient time at temperature for diffusion to take place between the core and the cover plates. Reductions of 60% or greater at temperatures varying from 1400–1800°F are required to bond titanium alloys. Rolling temperatures for other materials will vary with the material composition.

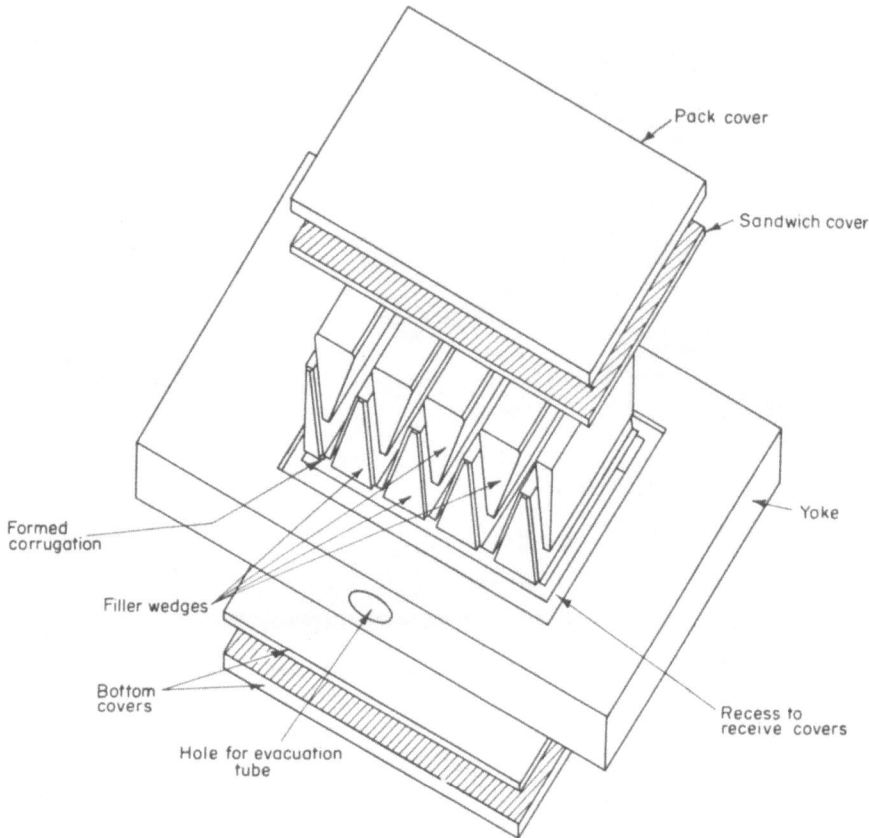

Fig. 4. Exploded view of pack assembly for producing roll-bonded corrugated-rib sandwich panels.

III. FORMING OPERATIONS

After rolling, and prior to removing the steel cover sheets and matrix material, the sandwich can be formed into various shapes in the same manner as a solid plate. Formability studies conducted on the Ti-6Al-4V truss-core configuration shown in Fig. 3a included bending, roll forming, and press forming.

Bend tests were conducted on $1\frac{1}{2}$ by 6 in. sections of a 0.25-in. Ti-6Al-4V panel complete with 0.10-in.-thick steel cover sheets. Total thickness of the panel was 0.450 in. Results of these tests showed that a minimum bend radius of 1T in the direction parallel to the ribs or 2T in the direction transverse to the ribs could be obtained without failure at temperatures of 1400°F or above. At

Fig. 5. Roll forming of a Ti-6Al-4V sandwich with steel matrix in place.

lower temperatures, cracks formed in the tension skin surfaces.

Roll-forming of Ti-6Al-4V panels to cylinder sections having a radius of 6 in. was done at 1600°F with the ribs in both the axial and circumferential directions. Figure 5 shows a section of Ti-6Al-4V panel being roll-formed into a cylinder, and Fig. 6 shows the cylinder after roll forming but prior to removal of the steel matrix.

Press forming of dome sections also was done at 1600°F as illustrated in Fig. 7. This photograph shows a dome section after forming and removal of the matrix material.

Removal of the Matrix Material

After forming to the desired shape, the filler material is removed from the sandwich by a leaching operation. For a titanium sandwich with a mild steel matrix, a boiling 50% nitric acid solution can be used. This solution does not attack titanium alloys and reacts rapidly with the mild steel matrix. It should be noted that the leaching step is used primarily to remove the matrix material from the interior of the sandwich. The steel cover plates are easily stripped off, and the yoke assembly cut off from the panel edges.

Fig. 6. Roll-formed cylinder of Ti-6Al-4V sandwich with steel matrix in place.

It is important that fresh acid be supplied to the acid–metal interface to maintain the maximum rate of leaching. This is done by circulating the acid through a manifold and acid-resistant tubing inserted into each of the channels. Otherwise the leaching rate decreases with time because of the depletion of the acid in the small channels. Table I summarizes the results of leaching studies conducted on 6-in.-long panels of the Ti-6Al-4V truss-core sandwich. In these tests, leaching was done from one end only. The other end was closed with an acid-resistant, stop-off material. It can be seen that mechanical vibration of the panel or aeration of the acid solution (accomplished by bubbling air through stainless steel tubes into the channels) greatly improved the leaching rate. However, the most effective method was the continuous circulation of fresh acid to the acid–steel interface.

After removal of the matrix material, the sandwich is essentially one integral unit, with a complete metallurgical bond between the core and the face sheets, as was shown in Fig. 2. Figure 8

Fig. 7. Press-formed dome of unalloyed-titanium sandwich after removal of the matrix
material.

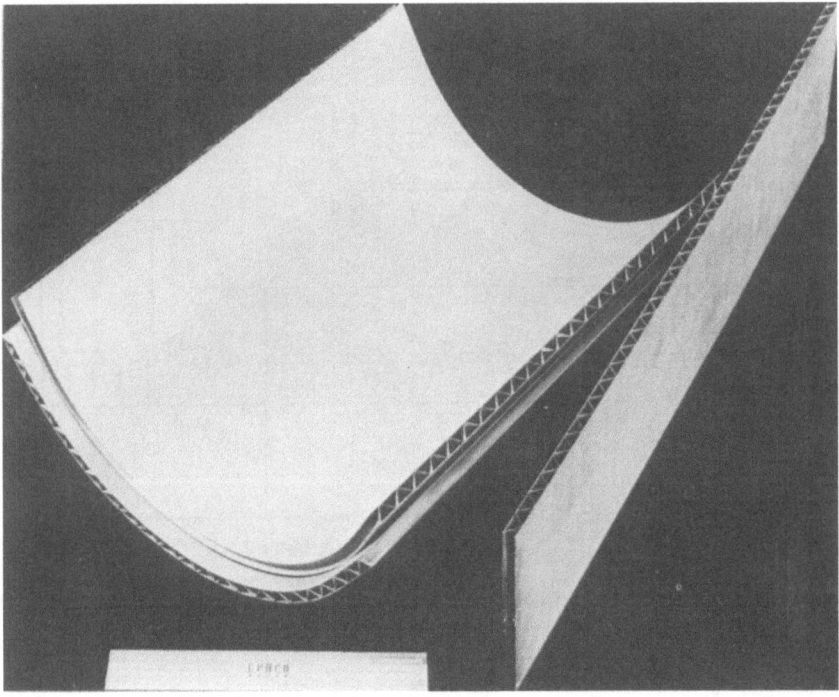

Fig. 8. Formed parts of Ti-6Al-4V sandwich after removal of the matrix material.

TABLE I

Leaching Rates of 1018 Steel Matrix from Ti-6Al-4V Truss-Core Sandwich in Boiling 50% HNO₃

Method	Average leaching rate for 6-in. channels, (in./hr)	Remarks
Stationary solution without agitation	3/4	Rate slows down rapidly with time
Mechanical vibration	1-1/8	"
Aeration of solution	1-5/8	"
Continuous circulation	2-1/8	Constant rate maintained

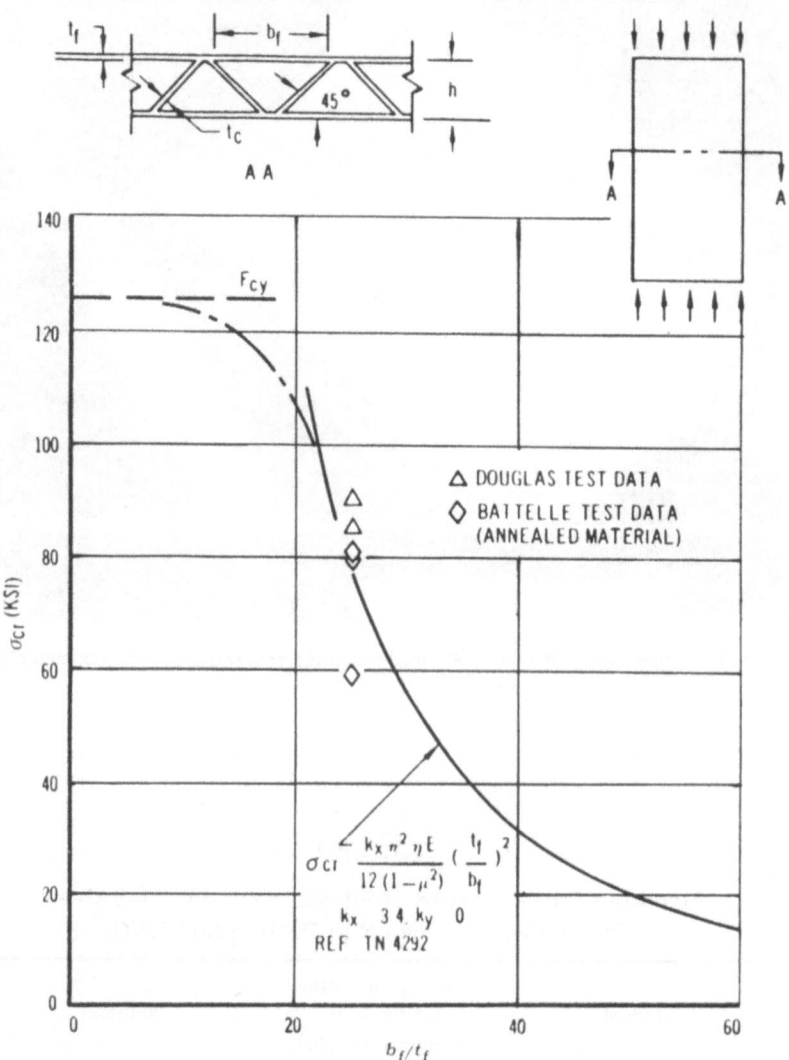

Fig. 9. Ti-6Al-4V roll-welded sandwich local instability compressive failure.

TABLE II

Tensile Properties of Ti-6Al-4V Sheet in the Annealed Condition*

Material	Specimen direction	Tensile strength, 1000 psi	Yield strength, 0.2% offset, 1000 psi	Elongation, % in 1 in.
As-received sheet	Longitudinal	144.5	137.5	16
As-received sheet	Transverse	144.5	137.5	12
From leached sandwich	Longitudinal	151.0	136.5	12
From leached sandwich	Transverse	146.0	137.5	11

*All values are the average of two tests.

shows some formed parts of Ti-6Al-4V sandwich after removal of the matrix material.

IV. MECHANICAL PROPERTIES

Base-Metal

With the intimate contact that occurs between the steel matrix and the Ti-6Al-4V sandwich during rolling, there is the possibility that contamination of the sandwich material may occur. Such contamination would be reflected in changes in mechanical properties. Tensile tests were conducted on a sample from the Ti-6Al-4V alloy sheet used in preparing roll-welded sandwich panels and on samples machined from the face sheets of fabricated and leached sandwich panels. The results given in Table II clearly show that there is no degradation of base metal properties that can be attributed to the roll-welding process.

Sandwich

In evaluation of the sandwich structure, consideration must be given in the geometry of the sandwich and the direction of the applied load. In stress situations where local instability may occur, such as axial compression, Anderson [1] has shown that the following equation can be used to predict buckling failures in truss-core sandwich structures:

$$\sigma_{cr} = \frac{K_x \pi^2 E}{12(1 - \mu^2)} \cdot \frac{(t_f)^2}{(b_f)}$$

TABLE III

Ti-6Al-4V Roll-Welded Sandwich
Compressive Local Instability
Test Results

Specimen number	Failure load, lb/in.	Equivalent thickness $\bar{t}$, in.	Failure stress,* psi
E1	3510	0.043	81,600
E2	3490	0.043	81,200
E3	2510	0.043	58,400
D1†	3790	0.0418	90,700
D2†	3700	0.0432	85,600

*All failures were by buckling; no failures occurred at the bond.
†Conducted at Douglas Aircraft Co.; other tests conducted at Battelle Memorial Institute.

Several tests have been conducted to verify the predictions of Anderson for buckling failures in the Ti-6Al-4V truss-core sandwich structure. Results are given in Table III and compared with Anderson's predicted failure in Fig. 9. They show good agreement between the theory and test for the geometry used in the test specimens.

Crack Propagation

The results of crack growth tests on a Ti-6Al-4V roll-welded truss-core sandwich are shown in Fig. 10. In this test, a starter notch in the form of a small hole was placed in one face. Cyclic stressing was started at a stress level of 48,000 psi until a small crack was initiated. The maximum stress level was then reduced to 29,000 psi and continued there until complete failure occurred on the face with the notch. At this time, a small crack was observed in the unnotched face and the test was discontinued. Post-test examination showed a crack growth pattern from the notched face to the unnotched face, as shown in Fig. 10.

In other tests with the crack growing parallel to the core direction, the crack growth characteristics are similar to a single sheet of the same material. However, because of the support of the other face through the core, the critical crack length is considerably larger than in an unsupported sheet.

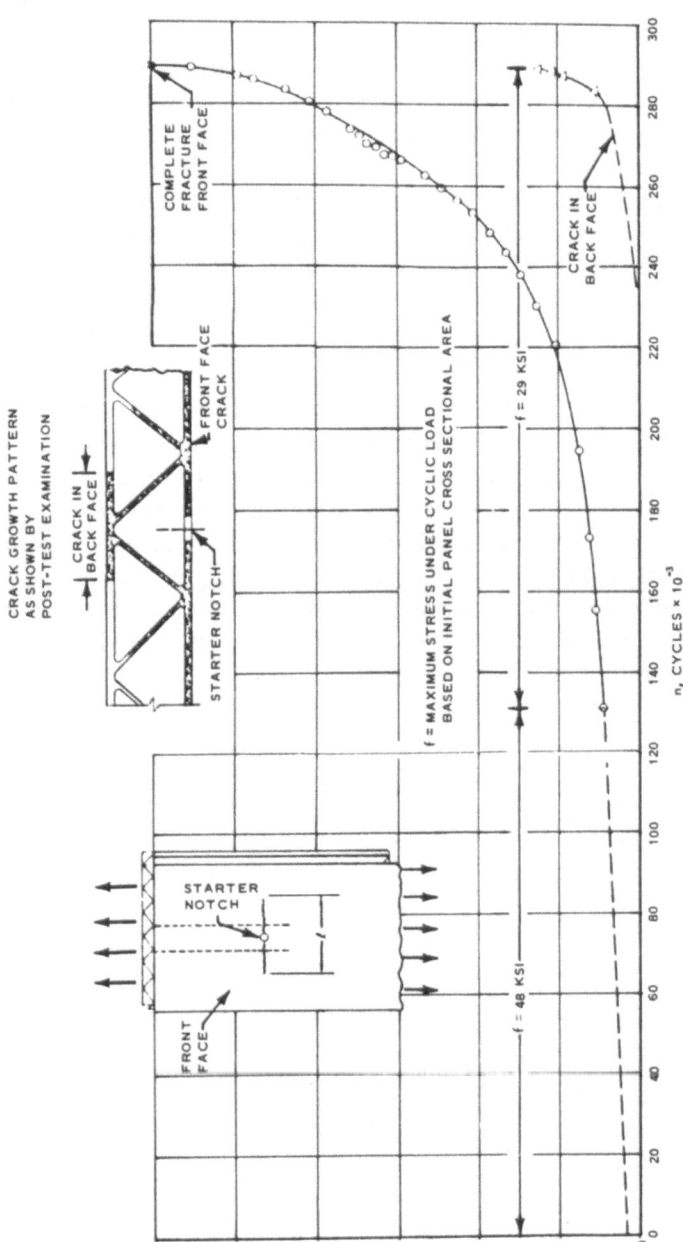

Fig. 10. Fatigue crack growth in titanium roll-welded sandwich.

Sonic Fatigue

An evaluation of the acoustic fatigue resistance of Ti-6Al-4V roll-welded, truss-core sandwich has been conducted in the progressive wave tube of the Douglas high-intensity sound system. This test subjected a small panel (~ 9 in. square) to an overall sound pressure of 162 to 164 dB for approximately 2 hr. Subsequent examination of the panel core-to-face sheet bonds with an ultrasonic C-scan recording instrument showed no damage at any point on the bond lines.

V. DISCUSSION

In the foregoing descriptions, the roll-welding process as applied to a Ti-6Al-4V alloy, truss-core sandwich has been emphasized. However, the process is not limited to titanium alloys, but may be used to produce sandwich structures or other composite structures from any material capable of being diffusion-bonded, and which will be inert to a chemical reagent that attacks the matrix material. For example, roll-welded structures have been produced in the laboratory from aluminum alloys, stainless steels, nickel-base alloys and refractory metal alloys. Dissimilar metals such as nickel and stainless steel also have been joined by roll-welding.

Since the major item of equipment required for the roll-welding of sandwich structures is a plate rolling mill, the size limitation for roll-welded sandwiches is the width of available rolling mills and size of auxiliary furnace capacity. To date, large laboratory-size

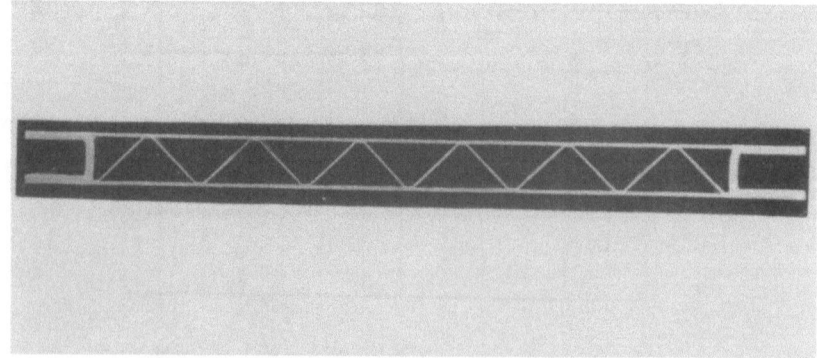

Fig. 11. Ti-6Al-4V alloy truss-core sandwich with roll-welded edge members.

panels measuring 3 by 6 ft have been produced. Obviously, much larger panels can be made by this process.

Another important feature of the process is the ability to roll-weld inserts, edge members, and even external fittings to the panel. By appropriately designing the initial pack, the final rolled panel will contain the desired attachments in the proper location. For example, Fig. 11 shows a cross-section of a laboratory panel containing edge members roll-welded into the panel which may be used for joining one panel to another.

V. CONCLUSION

It has been shown that with the roll-welding process it is possible to make large sandwich panels which can be formed to the shapes of various contours. The joints between the core and face sheets of the sandwich are metallurgically bonded in such a way that the sandwich structure is an integral unit. Because of the excellent bond, it is possible, then, to take advantage of the full potential of the properties of sandwich structures.

REFERENCE

1. M. S. Anderson, "Local Instability of the Elements of a Truss-Core Sandwich Plate," NASA TN4292 (July 1958).

Lightweight Aerospace Materials

E. P. Flint

Ipsen Industries, Inc., Rockford, Illinois

The primary objective of foamed materials development at Ipsen Industries, Inc., is to obtain refractory insulation for use at temperatures in excess of 4000°F in a vacuum or in other environments. The process used in preparing foams of refractory oxides and of metals depends on entrainment of air in a slurry of the powdered material, followed by drying and sintering. Firing in an oxidizing atmosphere is employed in manufacturing alumina and zirconia foams, which are in production at the company's subsidiary, Ipsen Ceramics, Inc., Pecatonica, Illinois. Thoria foams have been produced on a small scale.

Foams have also been made of molybdenum, stainless steel, Inconel, nickel, and tungsten by similar procedures, except that sintering must be carried out in a vacuum, hydrogen, or another atmosphere in order to reduce the oxide film that surrounds the particles.

Densities of the ceramic foams can be controlled from 10–50% of the theoretical values. Metal foams can be prepared in densities of 15–20% of the theoretical and higher. The foamed products, especially the metals, can be readily sawed and machined. Slight grinding of the foamed metal surfaces causes smearing and closing of the surface pores. Metal foams are readily brazed to solid metal in a vacuum furnace, and a strong composite is thereby obtained. Both ceramic and metal foams have many aerospace applications, the former for high-temperature insulation and structural components, and the latter for uses where reduction in weight and increase in resilience over the solid metal are important.

I. INTRODUCTION

The process developed for manufacturing refractory foams depends on the entrainment of air in a slurry of the refractory material, followed by drying and heat treating. Sintering in an oxidizing atmosphere is generally satisfactory for alumina, zirconia, and thoria, but many foamed materials must be heat-treated in a vacuum or in an inert atmosphere. With relatively few exceptions, this is true for metals, as well as for carbides, nitrides, silicides, and borides, since these materials are readily oxidized when heated in air.

II. PRODUCTION OF REFRACTORY FOAMS

Foamed alumina, zirconia, and another insulating brick for lower temperature service are currently in production. Firing of these materials is carried out in periodic or tunnel kilns, types of which are illustrated in Figs. 1 and 2. Normally the products are in the form of 9 by $4\frac{1}{2}$ by $2\frac{1}{2}$-in. brick, but many other shapes are made. A photograph of three kinds of our commercial insulating brick is shown in Fig. 3. The brick at the left of the photograph is designated as ICB-2600; the one at the center is foamed alumina, and, at the right, foamed zirconia. The ICB-2600 is currently being upgraded to a temperature capability of 2800°F and possibly 3000°F.

A photomacrograph of a section of foamed zirconia appears in Fig. 4. It will be noted that the walls of all of the cells are perforated with irregular openings which give a completely interconnected pore structure. This is important for a number of uses, one of which is in the insulating lining of Ipsen cold-wall vacuum furnaces where a rapid pump-down time is desired. Laboratory types of such furnaces are shown in Fig. 5. A production, zirconia-

Fig. 1. High-temperature elevator kiln.

Fig. 2. High-temperature tunnel kiln.

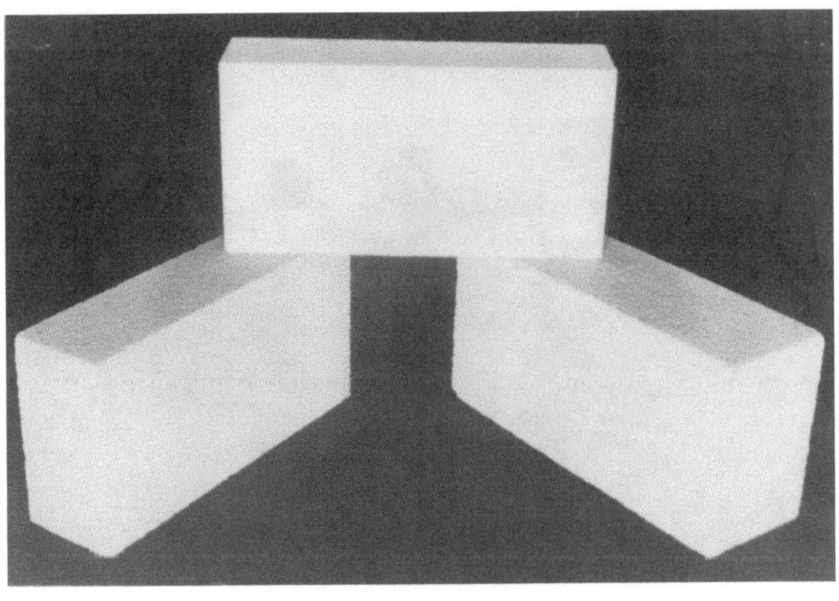

Fig. 3. Foamed ICB-2600 (left), alumina (middle), and zirconia brick.

Fig. 4. Foamed zirconia (20×).

Fig. 5. Laboratory, high-temperature vacuum furnaces.

lined vacuum furnace is illustrated in Fig. 6. The furnace is heated
with tungsten resistors and has a present operating limit of 4000°F
at a pressure of less than 1μ Hg.

III. TESTING OF REFRACTORY FOAMS

One of the more spectacular uses for foamed alumina and
zirconia has been in heat shields.   Both materials were in the
range of 10–12% of theoretical density.   Using its 14-in. hot gas
facility, Martin Company (Baltimore) tested the foamed alumina in
a nose-cone assembly 13 in. in diameter.   In one case, the alumina
modules of the assembly were impregnated with phenolic resin, and
in another assembly they were not.   The results showed that the
phenolic–alumina composite had excellent resistance to deteriora-
tion up to 3100°F, the maximum temperature of testing under the
severe conditions of thermal shock, combined with high dynamic

Fig. 6. Production-size 4000°F vacuum furnace.

Fig. 7. Phenolic-impregnated ceramic nose cap under test at Martin-Baltimore (Courtesy
Martin Company).

and sound pressure levels in this environment.   A photograph of
this nose cap under test conditions is shown in Fig. 7.

Small-scale foamed zirconia having 10–12% of theoretical density
was evaluated for thermal shock and erosion resistance. The speci-
men size used here was $2\frac{3}{4}$ in. in diameter and $1\frac{1}{4}$ in. thick. After
impregnation with phenolic resin, the disk was imbedded in a cast-
able plastic and then mounted in a water-cooled specimen holder.
After the start of the test, the sample was moved rapidly into the
hot gas stream and held in place for 15 min. This exposure produced a surface temperature of 3000°F, a 370-lb/ft^2 stagnation
pressure, and a 150-dB noise level, as well as highly erosive mass
flow and severe thermal shock. Examination of samples after test
showed that the plastic impregnant had decomposed at the surface of
the zirconia, but no erosion of the ceramic had occurred. In some
cases, a crack a few cell layers deep developed near the center of
the specimen.

At 10–12% of theoretical density, both alumina and zirconia,
when not impregnated with plastic, gave poor results in Martin's

Fig. 8. Thermal shock test stand (Ipsen).

hot gas facility. We have found that resistance to thermal shock of
foamed ceramic materials can be improved by increasing their
density somewhat and altering the distribution of bubbles in the materials, as well as by decreasing the coefficient of thermal expansion and increasing thermal conductivity.   However, intrinsic
properties cannot be altered when the finished product must be made
from a pure material.

IV. METHOD OF MEASURING THERMAL SHOCK RESISTANCE

Equipment for measuring the resistance of materials to thermal
shock is shown in Fig. 8. The apparatus is an oxyacetylene torch
consisting of a water-cooled head with twenty-five burner tips in a
2 by 2-in. area. Movement of the burner head toward the specimen
is controlled by means of a hydraulic cylinder which can be stopped
in any desired position.

The outer rows and the center tip of the burner are plugged, leaving only eight tips through which the flame is propagated. This permits even coverage by the flame of a $1\frac{1}{2}$ by $1\frac{1}{2}$ by $\frac{3}{4}$-in. specimen, which is surrounded by a refractory material to prevent edge effects. The specimen is weighed before insertion in the panel and mounted opposite the center of the flame. After lighting the torch, it is advanced by means of the hydraulic cylinder to a preset location, and the specimen is heated at a very nearly uniform rate of 50°F/sec to the desired temperature. On reaching the maximum temperature, the acetylene is turned off, and the specimen is allowed to stand in air. When cool, the test piece is removed from the panel, subjected to finger pressure to crumble spalled areas, and then weighed. The weight loss is taken as a measure of thermal shock resistance. This cycle of heating and cooling is repeated until the specimen is destroyed.

A different apparatus is used by the University of Dayton and has been applied to the testing of our materials. The equipment shown in Fig. 9 is an arc plasma torch into which is fed gaseous oxygen and nitrogen mixed in the percentages found in air, thus simulating re-entry conditions. A specimen $\frac{1}{2}$ by 2 by 2 in. is used, and the piece is held at a 45° angle to the nozzle of the burner, the flame of which is about 3 in. from the piece at its midpoint. The surface exposed to high heat is roughly $\frac{1}{2}$ in. in diameter. Temperatures at the hot face are read by an optical pyrometer, and the temperatures of the back face are followed by a thermocouple-recorder combination. A cycle of heating to 2250°F and cooling in air requires about 5 min.

Although these two methods are dissimilar in most respects, they have given qualitative agreement in eliminating pieces which are not resistant to thermal shock.

V. APPLICATION OF REFRACTORY FOAMS TO HIGH-TEMPERATURE COLD-WALL FURNACES

High thermal shock resistance is important not only in the development of materials for aerospace application, but also in the manufacture of furnaces for processing such materials. There are two types of cold-wall, resistance element furnaces: those which employ radiation shields and those which depend primarily on refractory insulation. Multiple radiation shields are less effective than a radiation shield used in conjunction with a porous refractory

Fig. 9. Thermal shock test stand (UDRI) (Courtesy University of Dayton).

or a structure built of porous refractory without any radiation shields for the following reasons:

1. Furnaces with refractory insulation require less power input than those equipped with heat shields.
2. With multiple radiation shields, it is difficult to prevent interaction between the metals used in the shields; for example, tungsten at the hot face, molybdenum at the next inner shield, and steel for the cooler shields. Buckling of the shields is almost impossible to prevent.
3. The radiation shield structure is relatively weak and therefore less adaptable for building large furnaces than the porous refractory.
4. Accidental leakage of air into a radiation shield furnace will damage or destroy molybdenum and tungsten shields very quickly, whereas refractory insulation is unaffected. Heating elements of the same metals are, of course, also damaged to some extent, but are less susceptible since they possess a greater cross section.

Cold-wall furnaces of both types are subject to contamination in the sintering process if volatiles have not been first eliminated in the dewaxing step.

Another type of cold-wall furnace is one in which a carbon resistance element is used as the heater and insulation is provided by finely divided carbon. These are satisfactory when carbon contamination of the work is not important. In any case, however, this type of furnace is troublesome; failures in the heater require frequent disassembly of the furnace, which is bothersome because of the dirty nature of the carbon black. Carbon-resistor furnaces are useful when reduction by carbon monoxide is desired, and provided that no deleterious reaction occurs between the carbon and the refractory insulation or between the carbon and the work.

The desired characteristics of refractory insulation for a high-temperature, cold-wall furnace with high production rate are as follows: (1) excellent thermal shock resistance, (2) low thermal conductivity, (3) low specific heat, (4) completely interconnecting pore structure, and (5) chemical stability at high temperatures and compatibility with other materials. Not all of these properties can be realized simultaneously. For example, low thermal conductivity reduces thermal shock resistance, and therefore compromises must be made to obtain the most suitable materials both for furnace linings and for aerospace applications.

VI. METAL FOAMS

The preceding discussion covers some of the aspects of our work on insulating high-temperature refractories and their applications. To the list of materials already mentioned should be added thoria which we have prepared in porous form. A photomacrograph of this material is shown in Fig. 10.

Investigation of methods of producing foamed metals was begun several years ago as an outgrowth of work on foamed refractories. This work was based on the premise that it would be simpler to produce sintered metal powder foams than to attempt foaming of a melt, since the latter procedure is extremely difficult to control.

Therefore, the method which we use is similar to that in foaming nonmetallics up to the point where the dried green foam is obtained. Subsequent heat treatment is different, since metal-to-metal bonding can be secured only by reduction of the oxide film which covers the surface of all reactive metal powders.

Figure 11 illustrates the partial pressure of oxygen from a number of metal oxides as a function of temperature. It can be seen that the oxides at the upper left in this figure, from copper to

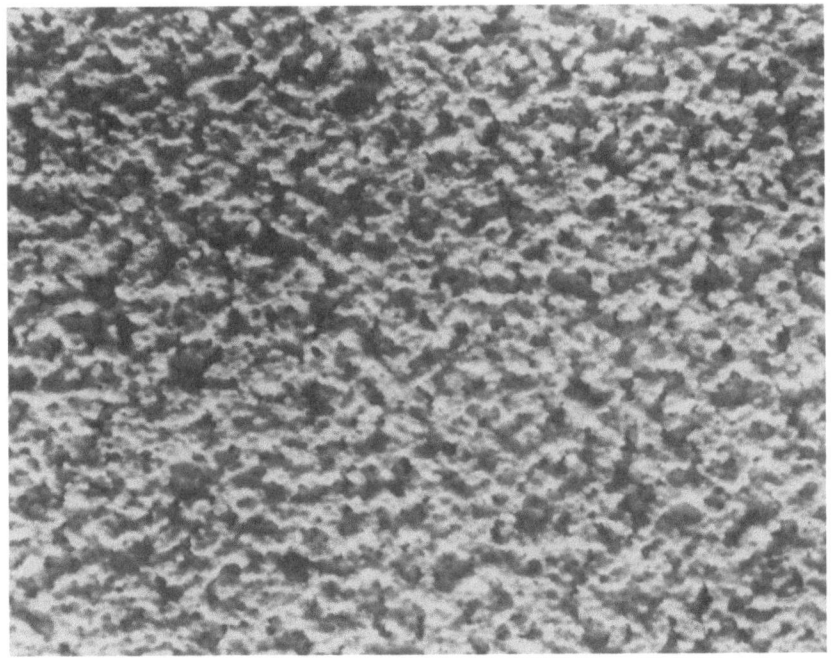

Fig. 10. Foamed thoria (54×).

tungsten, are easy to reduce because of a combination of high partial pressure and, in some cases, a high melting point of the metal. For the oxides at the lower right in this figure, reduction becomes difficult, and in some cases virtually impossible, because of the low partial pressure and limitations on allowable temperature of treatment. An example is the exceedingly great difficulty that is encountered in attempts to prepare a sintered aluminum foam.

In starting with aluminum powder, it is inevitable that all the particles that make up the dried green foam will be covered by a tightly adherent film of Al_2O_3. In conventional powder metallurgy, this presents no great problem since the film can be broken mechanically when the compact is pressed. No such possibility exists in preparing a foam, and the oxide film must be broken by other physical or chemical means. Since aluminum and aluminum oxide have melting points of 1220°F and 3720°F, respectively, sintering must be conducted at about 1200°F. At this temperature, according to Fig. 11, the oxygen partial pressure of aluminum oxide is approximately 10^{-47} torr.

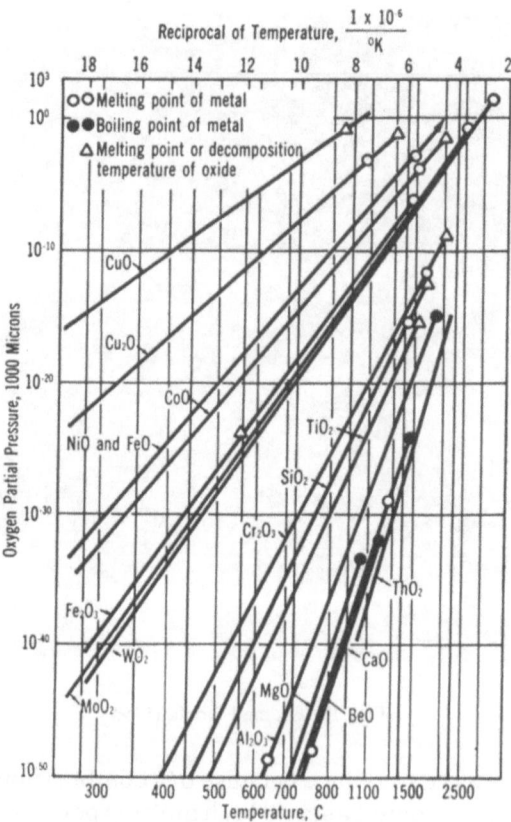

Fig. 11. Partial pressure of oxygen from metal oxides.

Reduction of alumina by hydrogen is governed by the reaction:

$$Al_2O_3 + 3H_2 = 2Al + 3H_2O \qquad \Delta H = -224,000 \text{ cal}$$

Because this reaction is so strongly endothermic, it has been found impossible to reduce alumina with hydrogen. Methods for sintering aluminum foam have been described in patents and other literature, but, to our knowledge, none has been put into commercial production.

The general method that we have developed in preparing sintered metal foams involves making the foam in the green state, followed by drying, and subsequent dewaxing at about 2300°F in a presintering furnace as shown in Fig. 12. This furnace is heated with a helical-wound Kanthal resistor and is lined with ICB-2600 brick. An over-all view of the furnace appears in Fig. 13.

Fig. 12. Vacuum presintering furnace for 2300°F operation.

Fig. 13. Overall view of vacuum presintering furnace.

Complete elimination of the organic materials is accomplished in the Kanthal furnace, and the volatiles are largely decomposed. The products are caught either in a large cold trap between the furnace and the mechanical pump or in the oil of the mechanical pump. The presintered pieces are strong enough to be transferred to another furnace for final sintering in atmosphere or vacuum.

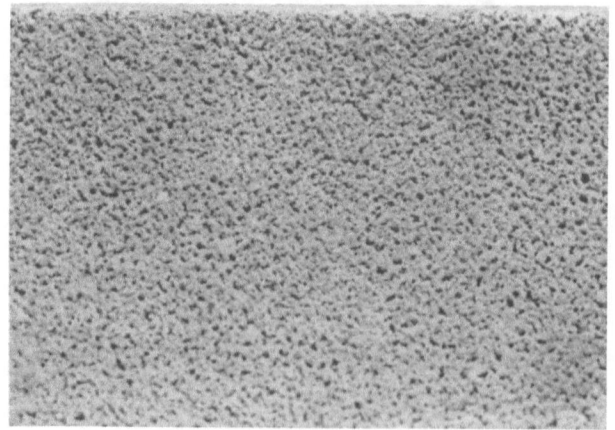

Fig. 14. Foamed Inconel slab.

Fig. 15. Foamed Inconel tensile test specimen.

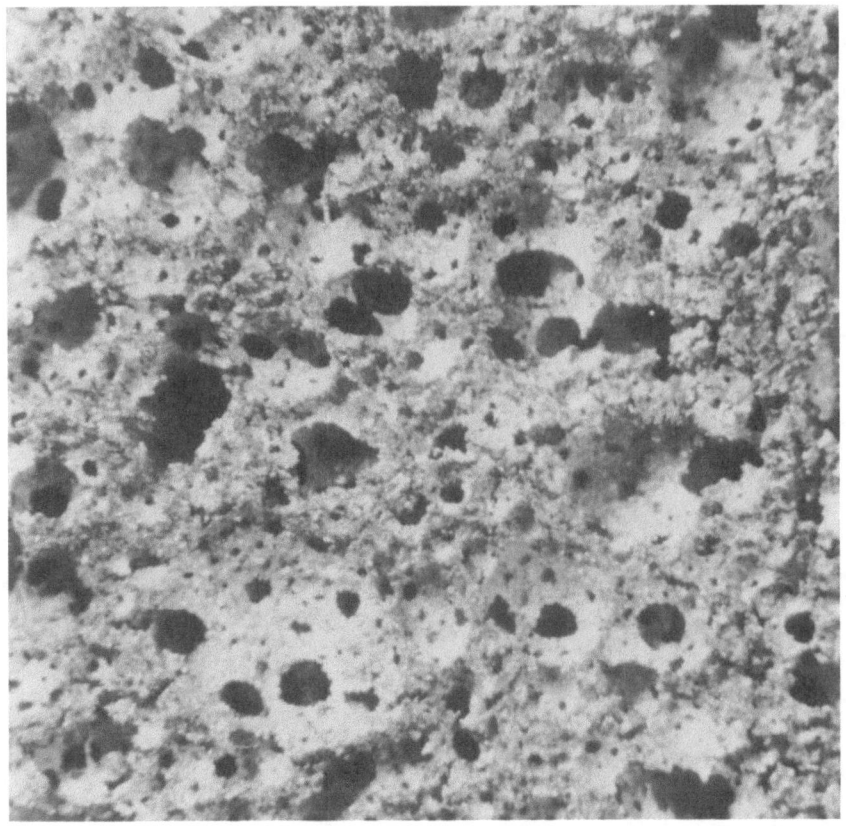

Fig. 16. Foamed molybdenum (16×).

Final sintering gives bright pieces of foamed metals of which the slab of Inconel shown in Fig. 14 is an example. Since the foamed Inconel is easily sawed, machined, and brazed, parts such as the tensile test specimen of foamed Inconel illustrated in Fig. 15 can be made.   Here the foamed Inconel is brazed to endpieces machined from wrought Inconel. Best results in joining are obtained when the porous metal is lightly ground with an abrasive wheel. Grinding seals the pores of the metal and prevents the brazing alloy from infiltrating the test specimen.   As in the case of the foamed ceramics, all pores in the foamed metals are completely inter-connecting.

The same technique can be applied to the preparation of foamed molybdenum. After the presintering step, the oxide film around the

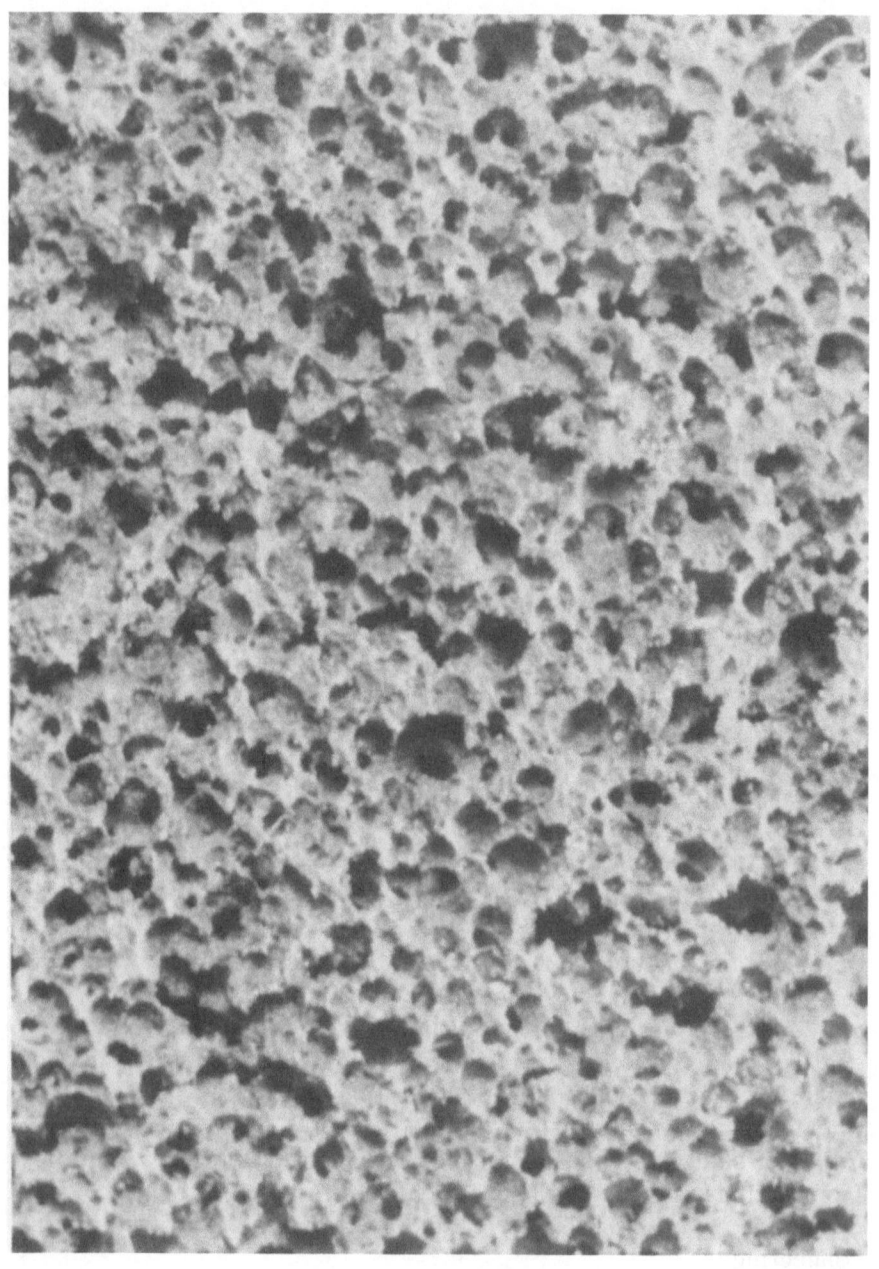

Fig. 17. Foamed tungsten (6.5×).

molybdenum particles can be easily reduced because of the instability in vacuum. The usual practice is to heat to 3900–4000°F for a suitable length of time. The foamed molybdenum obtained by this method is bright and strong (Fig. 16). Foamed tungsten is made similarly but, because of its high melting point (6170°F), a higher sintering temperature is required than in the preparation of foamed molybdenum. Figure 17 shows a photomacrograph of a sintered tungsten foam.

VII. CONCLUSIONS

In general, the sintered metal foams can be prepared in ranges of density from 15% to greater than 50% of theoretical. They have numerous aerospace applications as materials for transpiration cooling, energy absorption, and meteoroid and irradiation shields, to mention only a few. Industrial applications for metal foams are numerous and await large-scale production of these materials.

High-Temperature Materials and Coatings for the Aerospace Industry—Their Processing, Characteristics, and Applications (Part I)

R. W. Love, C. C. Esty, and W. M. Wheildon

Norton Company, Refractories Division, Worcester, Massachusetts

Ceramic materials are now coming into their own in the aerospace industry. Solid ceramics in combustion chambers and throats are replacing metals because of their extremely high fusion or decomposition temperatures. Coatings supplement both ceramic and metals by adding such properties as thermal insulation, erosion resistance, chemical resistance, and in the case of metals a temperature-resistant property.

Information is given on solid ceramics such as the carbides, borides, and oxides. This includes the relationship of manufacturing technique to such properties as thermal shock, erosion resistance, and strength. Other properties important to aerospace applications, such as oxidation resistance and melting point, are categorized. Recrystallized silicon carbide now presently used in aerospace systems is discussed.

Data are given on oxide coatings such as alumina, zirconia, chromia, and nickel oxide. This pertains to characteristics important to the aerospace industry, such as optical refractoriness and mechanical, electrical, and neutron shielding. Coating systems are explained in relationship to basic concepts and methods. Advantages of newer methods of ceramic coating techniques and their relationship to the aerospace industry are brought forth.

I. INTRODUCTION

Ceramic materials—oxides, carbides, nitrides, and borides—can in general be used at higher temperatures than most metals. Unfortunately, these materials do not take over where metals leave off; but, because at elevated temperatures they have properties not possessed by metals, they must be considered as construction materials for aerospace applications. However, in spite of the inherent weakness of ceramic materials as compared to metals at room temperature, ceramic materials with proper designs have been successfully used in certain space vehicle systems in a number of applications.

II. COATINGS

Fortunately, there is an alternative available to the spacecraft designer; use may be made of the desirable characteristics of

TABLE I

Melting Temperatures of Ceramics for Possible Spacecraft Use

Oxides	Carbides	Borides	Approx. melting point (°F)	Oxides	Carbides	Borides	Approx. melting point (°F)
	HfC		7000*	$ZrO_2 \cdot SiO_2$			4390†
	TaC		7000†	Y_2O_3			4370*
	ZrC		6350†		SiC		4350
	NbC		6320*	BeO			4260*
ThO_2			5970*	Gd_2O_3			4260
	TiC		5680†	$CaO \cdot ZrO_2$			4250†
		ZrB_2	5430†	La_2O_3			4180
		TaB_2	5430	Cr_2O_3			4110†
		NbB	5250		Be_3C		3900
UO_2			5220*	$MgO \cdot Al_2O_3$			3870†
	VC		5090	$MgO \cdot ZrO_2$			3850*
MgO			5070*	Al_2O_3			3840†
	Al_5C_3		5070			MoB_2	3810†
HfO_2			5030*			BaB_6	3810
	WC		5020†			CaB_6	3810
		CrB_2	5000†	NiO			3540†
CeO_2			4940*	BaO			3480
	MoC		4870	$2MgO \cdot SiO_2$			3420†
ZrO_2			4710†	TiO_2			3340*
		TiB_2	4710†	$3Al_2O_3 \cdot 2SiO_2$			3320†
CaO			4660	CoO			3280*
	B_4C		4440†	SiO_2			3140

*Has been experimentally flame-sprayed.
†Available as a flame-spray material.

current metals and alloys when their deficiencies are compensated for with a ceramic coating. Ceramic coatings are now used, not to substitute for metals, but to complement metal characteristics by adding further refractoriness, insulation, erosion resistance, oxidation and corrosion resistance, electrical resistance, or different optical characteristics.

One method of applying high melting point ceramic coatings to lower melting point substrates without excessive heating or other damage to the substrate is called flame-spraying. Table I lists some of the ceramic materials of interest as coatings to supplement substrate characteristics.

There are twenty-four oxides of interest between 3000–6000°F of which nineteen have been applied as flame-spray coatings, either experimentally or commercially. In addition, of the twelve carbides between 3900–7000°F, seven have been applied as coatings; and four of the eight borides between 3000–5500°F have been applied as coatings.

III. PROCESSES

The basic procedure of spraying a desired material in the form of molten or semimolten particles onto a cold metal surface as developed for metallizing is followed in the flame-spraying of ceramics. This technology has been expanded by new ideas and techniques which allow successful application of metal oxides by a number of simple and easily-employed commercial systems and by a variety of advanced experimental systems which will be discussed in detail later.

The basic systems for flame-spray application of ceramic coatings may be typed as follows: (1) powder in a combustion flame, (2) solid atomization in a combustion flame, (3) powder in a plasma flame, (4) solid atomization in a plasma flame, and (5) powder in a detonation flame.

In all these systems, a ceramic coating is created by a solid or powdered ceramic material which is heated in a flame, atomized, and projected, while still in the form of molten or semimolten particles, by a stream of gas moving at a high velocity onto the relatively cold surface of the article to be coated.

These impacting particles flatten, interlock, and overlap one another, so that they are securely bonded together to form a dense, coherent coating which is built up to the desired thickness. The

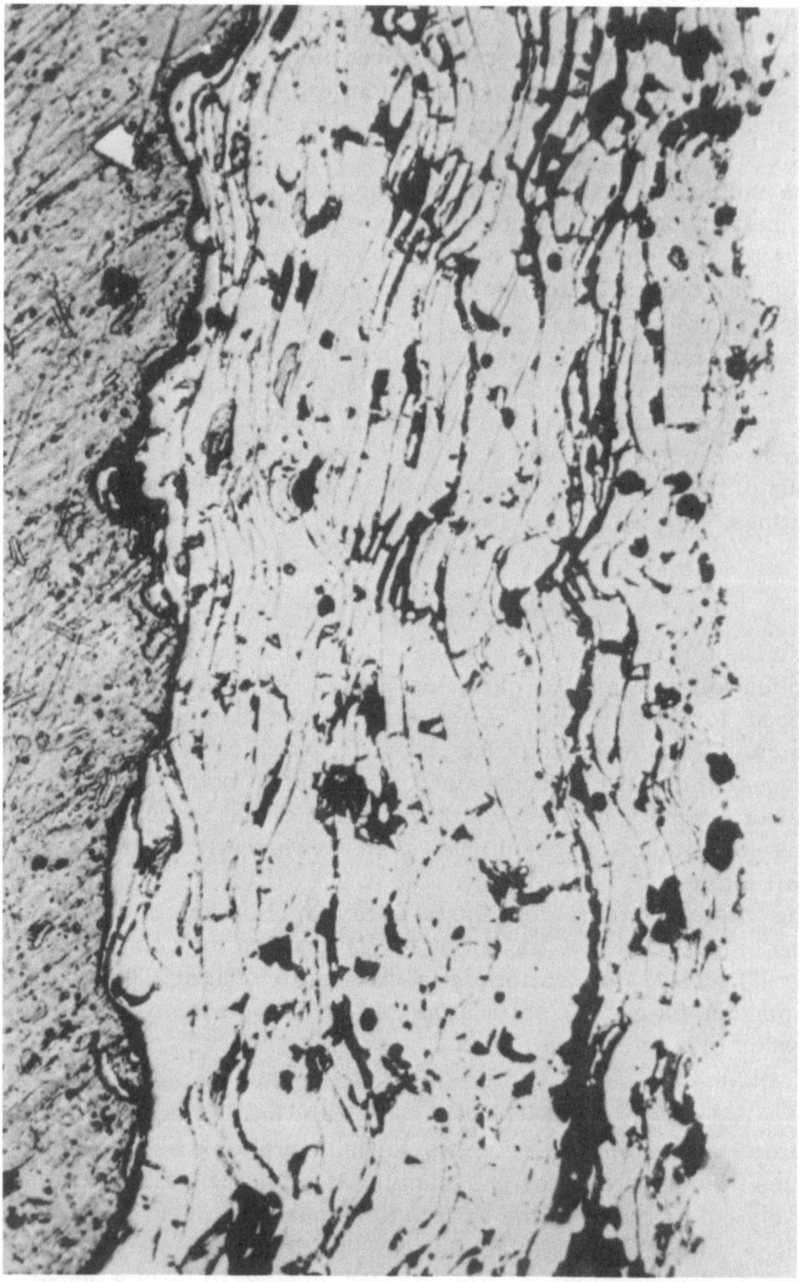

Fig. 1. Alumina rod-sprayed coating cross section (200×).

adherence of the coating to the surface results primarily from the mechanical fastening of the spray particles as they deform to take the shape of the suitably-prepared surface being coated. Thus, the proper degree of substrate roughness for anchoring is a prerequisite for creating an optimum coating.

Figure 1 is a photomicrograph in reflected light of a cross section of a type-2 system, rod-sprayed, alumina coating. The white area at the bottom is the metal substrate, the center area is the coating which is about 15 mil thick, and the top area is the resin in which the sample was mounted for polishing. The figure shows how the coating conforms to the roughened metal surface.

Much has been written on all the basic systems of flame-spraying, and each has its advantages and disadvantages. The authors are primarily, though not solely, concerned with solid atomization systems which are currently known as Rokide processes. Therefore, to avoid repetition of available work, this paper will deal, for the main part, with the newer developments and aspects of the solid atomization or Rokide process.

The solid atomization process differs from the powder process in that a solid body, usually in the form of a rod, is used as the coating material source. This is fed into a high-temperature flame where the end of the rod is melted, the molten material is disintegrated by a rapid flow of compressed gas, and the resulting spray is delivered by the high-velocity flow of gases to the substrate. Figure 2 shows schematically the system for solid atomization employing a ceramic rod of the desired coating material.

Such a system permits the use of a very short flame or heat zone, since the end of the rod must be molten in order to provide a liquid film that will separate from the rod and atomize to the desired particle size. Also, a close control of spraying rate is achieved by controlling the rate of feed of the rod into the flame.

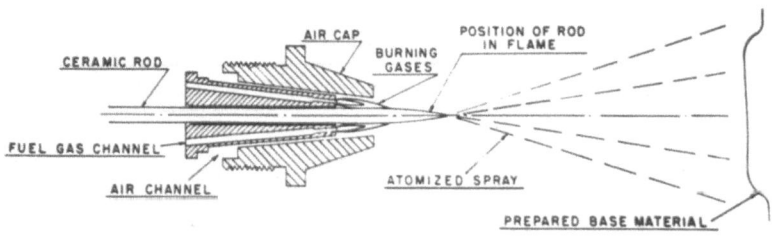

Fig. 2. Rod-spray atomization system.

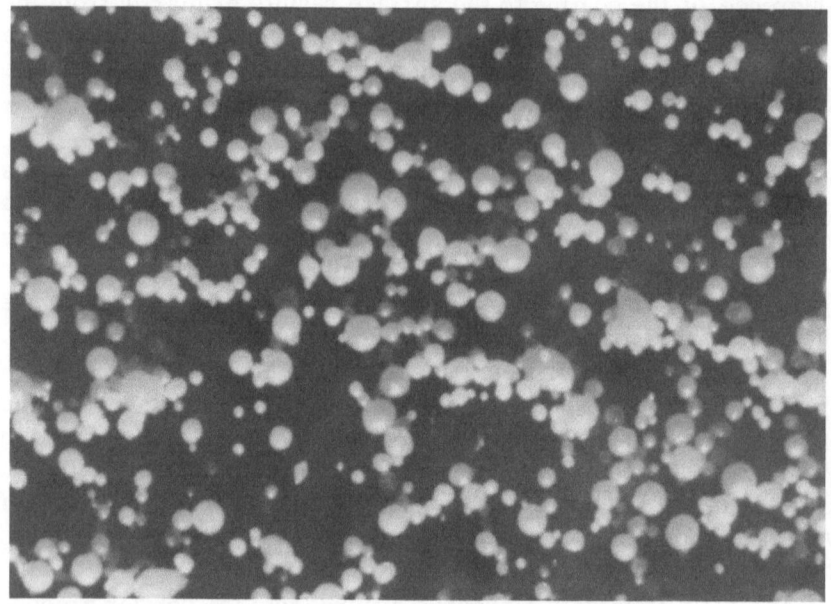

Fig. 3. Zirconium dioxide spheres.

IV. CHARACTERISTICS OF THE COATING

The particle size obtained is essentially independent of the particle size in the rod; the size of the sprayed particles is determined by the fluidity of the material and the velocity of the impelling compressed-gas stream. Particle sizes ranging from 1–150 μ have been observed. Figure 3 shows spheres of zirconium dioxide produced by flame-spraying $\frac{1}{4}$-in.-diameter zirconia rods in an oxyacetylene flame. They average about 75 μ and produce an optimum coating.

TABLE II

Process	Maximum spray particle velocity (ft/sec)
Detonation (Linde)	2400
Rokide (Solid Atomization)	600
Plasma (Powder)	450
Combustion Flame (Powder)	150

Actual spray particle velocities have been measured by the National Bureau of Standards. Table II shows the maximum value of the particle velocity for this process compared with the maximum reported velocities for the other various ceramic spray systems.

Obviously, particles with high velocity and low viscosity will flatten to a great extent on striking a surface. Particles such as these force themselves into the irregularities of the surface which they strike, bond securely, and tend to have few pores. Conversely, particles with low velocity deform only slightly and produce a porous coating.

The total porosity of coatings made from well-fused particles at high velocities is less than 10%. Considering the very slight heating of the substrate to which coatings are applied, it is surprising that densities above 90% of theoretical can be easily attained. Such densities are often difficult to obtain in conventional monolithic ceramic bodies. Proper control of the detonation process can produce alumina coatings of a total porosity as low as 1%; such coatings, however, are liable to suffer under severe thermal-shock conditions.

After the creation of a quality coating, it becomes most important to improve rates of coating application and control of spray pattern. This can be done by equipment development or spray material development or both.

V. ADVANCED SYSTEMS

New combustion-materials systems and new plasma-materials systems are being developed to advance flame-spray coating practice.

Figure 4 shows schematically a fuel gas injector system, currently in various stages of field and limited commercial use, that shows a potential advantage of some 50% or more in ratio of coating laydown with full maintenance of coating quality.

As shown in Fig. 2, fuel gases are normally premixed before entering the burning nozzle. The energy applied to melt the rod is then controlled by the amount of gas available to be burned, which is in turn limited by allowable acetylene pressures and gas orifice area controlled to avoid backfiring.

The experimental atomizing system in Fig. 4 does not premix the fuel gases, but brings the oxygen and acetylene separately and

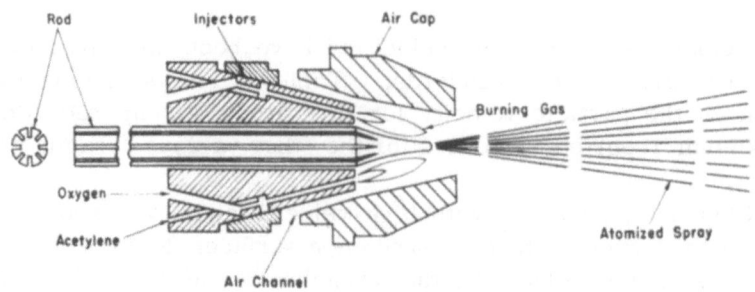

Fig. 4. Advanced rod-spray system.

directly to the burning nozzle. The oxygen, which is not limited in
allowable pressure, is used to inject the acetylene directly into
the burning orifice. Thus, as each small burning nozzle has ten
gas orifices, there are in effect ten gas injector systems. These
provide a higher volume of fuel gas through the same size orifices,
thus burning with consequently greater energy in order to melt the
rod. This type of injector nozzle coupled with the compound-angle
air cap has raised the rate of spraying, and thus has substantially
reduced spraying cost.

While this system now appears simple and quite obvious, the
selection of the right orifice sizes and proper angles, and consideration of the relationship between the two proved a rather complicated matter. More than three hundred configurations, geometries, and dimensions were tried before reaching the above
solution. As yet, this system has been applied only to conventional,
single-nozzle-type guns. A prototype for future equipment is the
fully automated multinozzled automatic high-production Rokide unit
(Fig. 5). It is presently under construction, and individual prototype parts have been laboratory-checked. This unit is expected
not only to speed up coating application, but also, at the same time,
to simplify and control spray handling and material feed. A built-in
versatility will allow substitution of different heat-source heads as
required.

In addition to equipment development, the make-up of the rods
has proved to be of utmost importance in gaining desired spraying
characteristics. The proper chemical composition and geometry
contribute greatly toward the achievement of a good performing
coating. Because most ceramics are good thermal insulators, a
large surface area for flame impingement is desirable. A new
fluted configuration which gives a 100% increase in surface area with

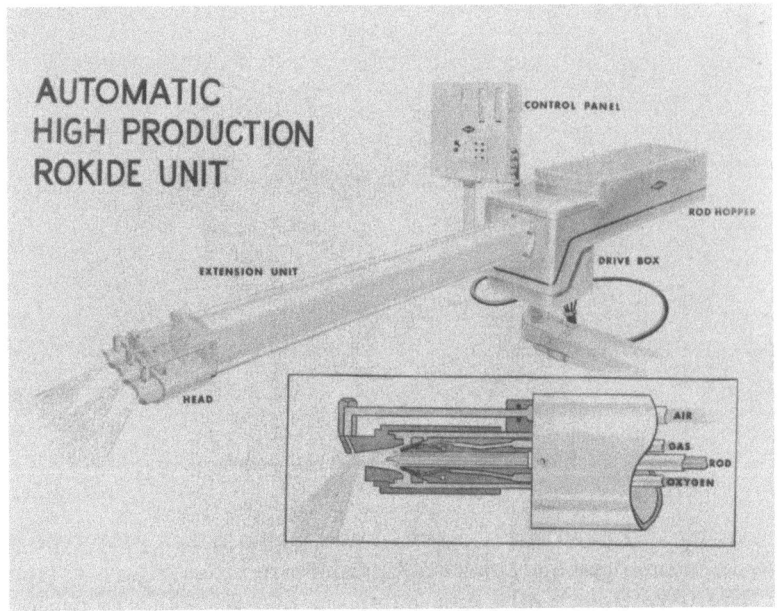

Fig. 5. Automatic high-production Rokide unit.

very little loss in rod volume (about 8%) because of the very narrow slots is shown in Fig. 4. It has several advantages—faster spraying rate, less susceptibility to thermal shock, and a tendency to weld successive rods into a continuous train. For the moment, this system appears to be about the optimum that can be expected from solid atomization employing combustion gases. A variety of gases and finely-powdered metals have been considered and experimented with as fuels, but none have proved as practical as oxyacetylene.

However, the recent practical development of the arc plasma flame torch for spraying is promising. Experimental indications are that with spraying of ceramic powders, such as zirconia, high rates of coating laydown are limited only by the power input into the gun and the ability to cool the substrate. It now appears that $4-4\frac{1}{2}$ kW of power is required for each pound of zirconia coating applied, and guns sufficiently sized to handle greater than 10 lb/hr are now readily available.

Because of the possible advantages of plasma as a heat source for the solid atomization process, a rod-plasma gun is being created which is expected to affect the economics of coating application and produce superior or different wanted characteristics.

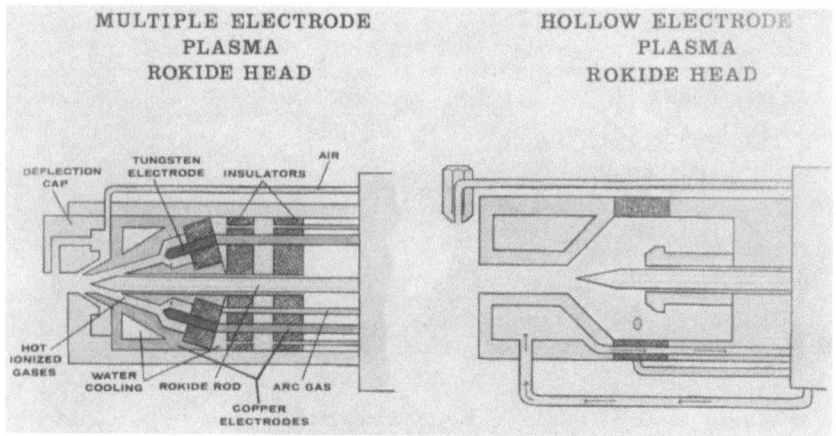

Fig. 6. Experimental rod plasma heads.

During the past three years, more than a dozen prototype guns have been engineered, constructed, and tried involving a variety of arc configurations, different numbers of arcs, and variable rod feed angles. Feeding a rod directly into or through an arc or plurality of arcs was not successful because the rod either volatilized or separated into large molten particles unsuitable for coating. Therefore, the approach was changed to feed the rod into the plasma gas flame or hot stream of gas heated by the arc. This has been much more successful in melting the rod congruently. The plasma flame is restricted around the rod in such a manner that heat transfer to the rod is maximized and sufficient gas velocity is obtained to remove continually a liquid melt film as a fine spray.

Figure 6 shows schematically two of the experimental systems intended for attachment directly to the high volume drive unit of Fig. 5. Of the two systems shown, the multiple-electrode head is showing the most promise. Because this unit has just recently been completed, no test data have been assembled. However, in preliminary tests of coating rates of zirconia in a single-electrode prototype gun with the rod entering the plasma flame at a similar angle, indicated a requirement of 6–7 kW/lb of coating applied. Experimental guns have operated easily in 6-lb/hr laydown rates, and the more recent designs are expected to do even better. Previous combustion systems obtained a maximum laydown rate of only 2 lb/hr.

This briefly summarizes the basic concepts of flame-spraying, certain methods of application, some of the technological improvements, and some plans for future systems.

VI. APPLICATIONS

Some of the available coating materials and their functions for space technology applications will now be examined and illustrated.

Because of the simplicity of the process, coatings can be prepared from a wide variety of materials. The majority of the investigations concerning these coatings have dealt with simple compounds, but complex mixtures such as cermets and porcelains will also produce coatings. Two limitations which determine the materials that will form coatings are: (1) the material must melt in the flame and must not volatilize; and (2) the molten material must spray as drops rather than fibers.

For example, nickel oxide or boron carbide rods melt in the flame, project as spheres, and freeze on the surface as coatings of these same materials. On the other hand, silicon carbide rods do not melt, but dissociate when an attempt is made to spray them. A coating of silica slowly builds on a surface in the path of the spray pattern.

An example of the second limitation is illustrated by attempts to spray rods of pure fused silica. When a rod of fused silica is fed through the flame, the silica softens, and the stream of rapidly moving gas produces fibers, rather than drops of liquid silica. A mass of fibers, rather than a coating, is obtained.

With these limitations in mind, the authors have successfully applied experimental flame-spraying coatings with more than fifty materials and combinations of materials. Some of the more interesting of these are listed in Tables III–VI together with their physical, mechanical, electrical, thermal, and chemical properties.

The aerospace industry in general employs coatings with one or more of the following types of characteristics: optical, refractory, electrical, mechanical, and nuclear shielding.

Optical

The choice of materials for the outer surfaces of spacecraft is determined primarily by the need for temperature control and stability of the material in vacuum within limits established by the requirements of the instrument payload. The surface temperature of a craft in space is controlled by the optical characteristics of

TABLE III

Physical Properties of Rod-Sprayed Coatings

Types of coatings	Color	Crystal form	Bulk density (average g/cc)	Crystal hardness (Knoop scale monolithic crystal)	Porosity	Permeability
Rokide A aluminum oxide	White	Gamma type	3.3	2000	8 (7% open)	Slight
Rokide Z zirconium oxide	Light tan	Cubic	5.2	1000	8 (7% open)	Slight
Rokide ZS zirconium silicate	Light tan	Cubic ZrO_2 in siliceous glass	3.8	1000	8 (4% open)	Slight
Rokide C chrome oxide	Black	Hexagonal	4.6	1900	4 (2% open)	Very slight
Rokide MA magnesium aluminate	White	Cubic	3.3		6 (4% open)	Slight
Stainless steel 304			8	400 approx.	0	None

the surface and by the external and internal energy arriving at the surface. In other words, surface materials are selected to have proper characteristics for absorbing, reflecting, and emitting radiation in required amounts. In addition, the temperature control surface must have sufficient durability to withstand space environment without change in radiation characteristics.

A tremendous amount of investigation and data collecting has been performed on this subject, but it suffices here to state that pure oxide flame-spray Rokide coatings can be prepared covering a wide range of optical characteristics and have performed satisfactorily in space over substantial periods of time.

Figure 7 shows a range of color gradations available in different experimental and standard Rokide coatings. Certain of these proved ideally suited on Explorer, Tiros, and Pioneer satellites.

Figure 8 shows the Explorer I, first of the U. S. series of satellites. Requirements of the electronic payload imposed temperature limits of 23–113°F. Thus, as the satellite travels in space, the skin must absorb some sunlight and yet not give off too much heat while in the shadow of the earth. Jet Propulsion Laboratory's calculation employing the known optical characteristics indicated that solid atomization flame-sprayed alumina in eight strips, 0.625 in. wide by 0.005 in. thick, placed lengthwise along the outside of the Explorer's instrumentation cylinder and continuing onto the nose cone would do the job. The job was done successfully as telemetering information recorded temperature control between 50 and 86°F.

Figure 9 shows a sketch representative of the several Tiros meteorological satellites flown to date. These vehicles employed area coverage of alumina coating similar to that indicated for the Explorer vehicles. Successful operation of these vehicles in orbits exceeding 400 miles for many hundred trips has provided many thousands of cloud-cover photos which are interpreted into useful weather data. While the coating had a similar purpose to the coating on the Explorer craft, it was applied in an entirely different manner. No spraying was done directly on the vehicle; the coating was prepared as a free-standing body in the shapes shown in the background of Fig. 9 and in individual monolithic coating shapes cemented to the vehicle surface. The individual coating shapes were produced by spraying onto a metal template coated with salt, then dissolving out the salt in water to free the thin (0.010 in.) coating sections.

TABLE IV

Mechanical Properties of Rod-Sprayed Coatings

Types of coatings	Compressive strength (psi at ambient temp.)	Adherence to steel (approx. psi)	Profilometer surface finish — rms range (μ in.)	Abrasion resistance (rubbing wear impact abras.)	Coefficient of friction	Strain % elongation per unit of length (0.020-in.-thick coating)	Vibration (cycles)
Rokide A aluminum oxide	37,000	Steel 1000 Non Ferrous 600	As coated 200–300 As ground 30–50 As lapped 25–45	Very good	0.10 Against 440-C SS	0.7	200,000,000 cycles without failure (15,000–16,000 psi cyclic stress)
Rokide Z zirconium oxide	21,000	Steel 1000 Non Ferrous 600	As coated 200–300 As ground 30–50 As lapped 25–45	Good	0.10 Against 440-C SS	1.4	

Material					
Rokide ZS zirconium silicate		Steel 1000 Non Ferrous 600 As coated 200−300 As ground 30−50 As lapped 25−45	Good	0.10 Against 440-C SS	0.7
Rokide C chrome oxide	105 000	As coated 200−300 As ground 15−25 As lapped 10−20	Very good	0.11 Against brass	1.3
Rokide MA magnesium aluminate		As coated 200−300 As ground 30−50 As lapped 25−45	Good		
Stainless steel [304]	70 000−100 000			0.2	

TABLE V

Electrical Properties of Rod-Sprayed Coatings

Types of coatings	Conductivity	Electrical resistivity (ohm-in.)	Dielectric strength (AC volts per mil thick)	Dielectric constant (at $0-600°F$)	Dissipation factor (at $0-600°F$)	Loss factor
Rokide A aluminum oxide	Nonconductor	4.5×10^6 at 500°F 1.2×10^5 at 800°F	0.005-in. thick - 160 0.010-in. thick - 120 0.020-in. thick - 65 0.030-in. thick - 48 +	10	0.05	0.5
Rokide Z zirconium oxide	Nonconductor (amb. temp.) Increase rapidly at 1200°C	2.7×10^4 at 500°F 2.1×10^2 at 800°F		35	0.15	5.25
Rokide ZS zirconium silicate	Nonconductor	1.1×10^6 at 500°F 4×10^2 at 800°F		15	0.08	1.20
Rokide C chrome oxide	Nonconductor					
Rokide MA magnesium aluminate	Nonconductor					
Stainless steel 304	Conductor	28.3×10^{-6} at 68°F 45.6×10^{-6} at 1200°F				

TABLE VI

Thermal and Chemical Properties of Rod–Sprayed Coatings

Types of coatings	Melt. temperature (°F)	Mean specific heat (Btu/lb/°F)	Coefficient of expansion (in./in./°F)	Conductivity (Btu/hr/ft²-in./°F coating only)	Total emittance	Thermal shock resistance	Chemical Composition (%)	Chemical Resistance to acids, alkalis†
Rokide A aluminum oxide	3600	0.28 (90°–3100°F)	4.1×10^{-6} (70°–2250°F)	19 (1000–2000°F)	0.8–0.4 (200–1000°C)	Good	Pure aluminum oxide 98.6 Al_2O_3	Good Good (except hot)†
Rokide Z zirconium oxide	4500	0.175 (80°–2550°F)	5.4×10^{-6} (70°–2250°F)	8 (1000–2000°F)	0.7–0.3 (200–1000°C)	Very good	ZrO_2 + 2 wt. % HfO_2 + 3.5–5.0 wt. % CaO	Good (except HF) Good†
Rokide ZS zirconium silicate	3000	0.15 est	4.2×10^{-6} (70°–1100°F)	15 (1000–2000°F)	0.7–0.3 (200–1000°C)	Good	65 ZrO_2 34 SiO_2	Fair (except HF) Good† (except hot)
Rokide C chrome oxide	3000	0.2 (60°–2700°F)	5.0×10^{-6} (70°–2000°F)	18 (est) (1000–2000°F)	0.8–0.9 (100–1200°C)	Moderate	85 Cr_2O_3	Good (except HF) Good† (except hot)
Rokide MA magnesium aluminate	3500	0.25 (70°–1832°F)	4.5×10^{-6} (70°–2000°F)	18	0.7–0.3 (100–1200°C)	Good	98 MgO - Al_2O_3	Good (except HF) Good† (except hot)
Stainless steel 304	2600	0.12 (32°–212°F)	10.5–11.5×10^{-6} (32°–1800°F)	150 (1000°F)	0.45 (as rolled)			Good (except hot) Good†

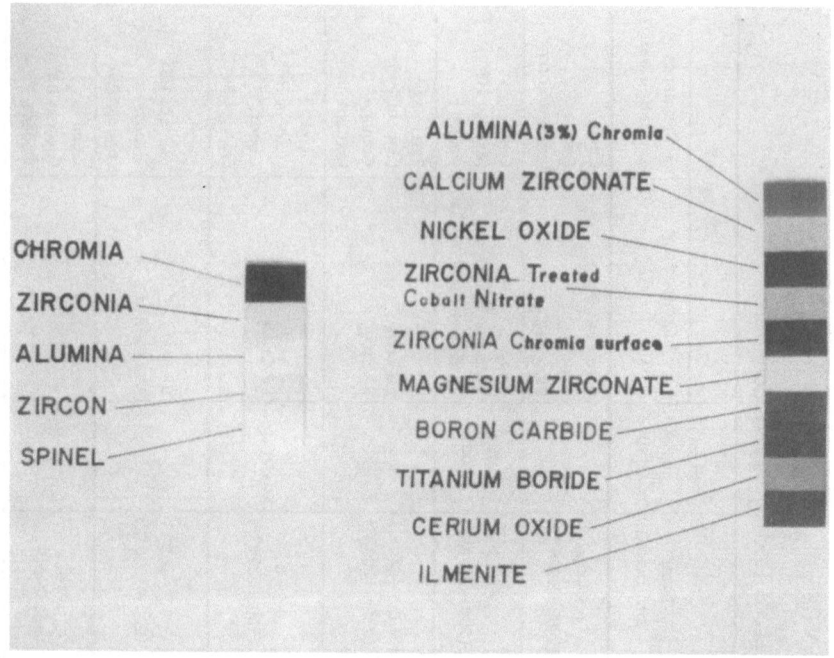

Fig. 7. Color range of ceramic coatings.

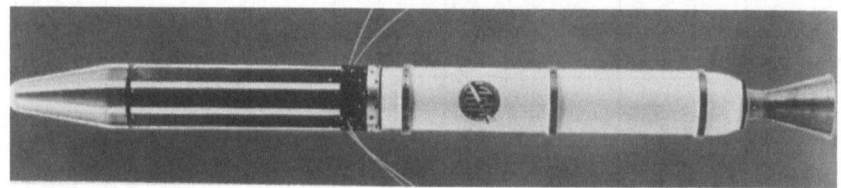

Fig. 8. Explorer I satellite.

Both of these illustrations have taken advantage of the stable reflectance characteristics of aluminum oxide.

For applications requiring absorption characteristics, chromium oxide and nickel oxide are entering the picture. Recent measurements of chromium oxide by a number of investigators show it to be particularly stable in the high emissivity range (0.8–0.9) under a variety of conditions and thicknesses. One group is considering it as a standard.*

*For those particularly interested in emittance and absorption characteristics of flame-sprayed rod coatings, considerable data have been assembled by Norton Company and are available through the authors of this work.

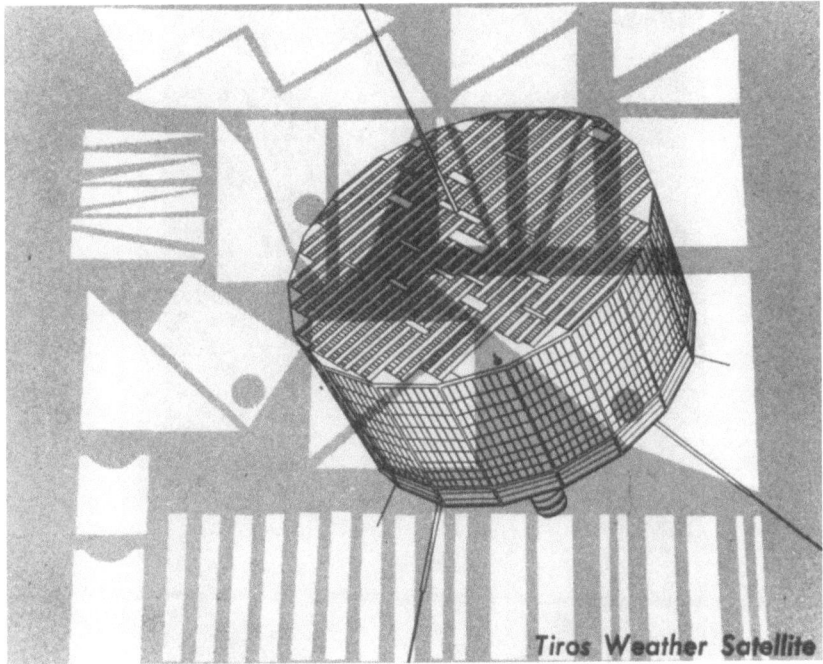

Fig. 9. Tiros weather satellite.

Refractory

Refractory applications of flame-sprayed coatings are typified by protection supplied to thrust chamber nozzles of spacecraft power plants.

Figure 10 shows the Bell Aerosystem's Agena rocket engine, a reliable propulsion unit for a steadily increasing family of space vehicles including the Ranger, Mariner, and Gemini series. A 0.010-in. thickness of aluminum oxide (rod-sprayed) is employed on the inside diameter of the thrust cone to form a temperature and erosion barrier between the hot gas stream and the thin metal substrate. Under experimental and nonequilibrium conditions, temperature drops of 1500°F have been recorded through this thickness of coating.

Figure 11 shows the thrust nozzle of the famous Bull Pup missile together with a photo of a Navy plane carrying this destructive air-to-ground missile. On the business-end of this dependable power plant, a copper nozzle is protected by a thin zirconium oxide coating. Many are aware of the potentials of this missile, but few

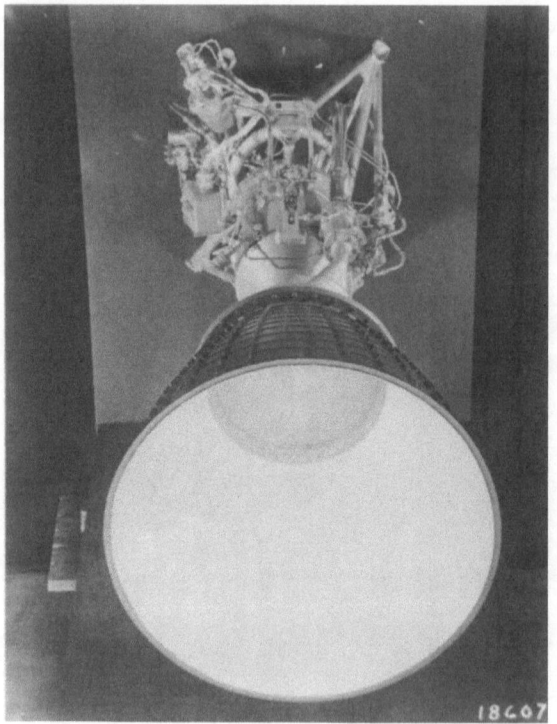

Fig. 10. Agena rocket engine.

realize that coatings, a matter of thousandths of inches in thickness, must withstand erosive gases of a combusting liquid propellant at about 4000°F.

Electrical

In the electrical field in space, flame-sprayed coatings have found a place in the insulating and mounting of electronic and electrical components such as temperature-sensing elements, strain gages, and small motor armatures.

Three characteristics of the coatings have contributed to success in these fields: (1) The structure of the flame-sprayed coating is such that it will withstand, without cracking, strains greater than those induced by expansion, contraction, or flexure of the backing material. (2) Coating can be sprayed around wires to fix them into a predetermined position. (3) Alumina (Al_2O_3) provides excellent electrical insulation at high temperatures.

Fig. 11. Navy Bull Pup missiles.

Fig. 12. Small motor armature — alumina shaft insulation.

Figure 12 illustrates one of the newer electrical applications for sprayed alumina coatings. It is used here to insulate the small motor armature from the drive shaft in order to avoid grounding problems through the shaft. At the same time, the coating must be sufficiently rugged to drive the shaft at elevated temperatures. A 0.040-in.-thick coating does the job by withstanding 2500 V breakdown test. The coating is rugged enough following diamond grinding to accommodate a press fit into the armature assembly.

Mechanical

One of the more interesting mechanical applications for sprayed coatings, other than a positioning and fastening medium as described under electrical application, is its use for mechanical seals.

Flame-sprayed coatings are capable of being finished by grinding as low as 7–10 rms and flatness readings of less than three light bands. In such a condition, zircon flame-sprayed applied coatings provide a particularly compatible mating surface with graphite.

Figure 13 shows a commercial mechanical seal employing a rod-sprayed zircon coating against a graphite monolithic counterpart.

Nuclear Shielding

The fifth type of coating characteristic which is expected to be of growing interest in the space technology field is nuclear shielding.

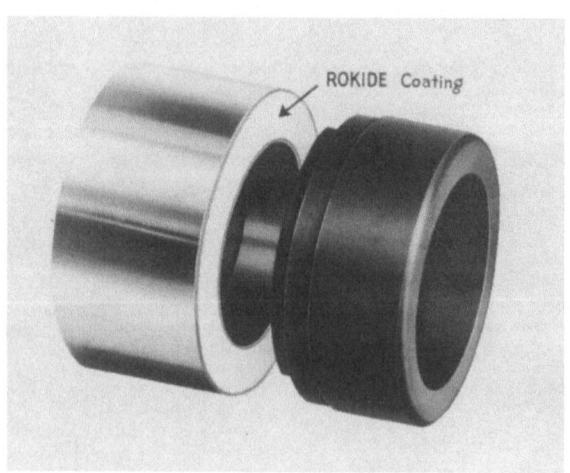

Fig. 13. Mechanical seal.

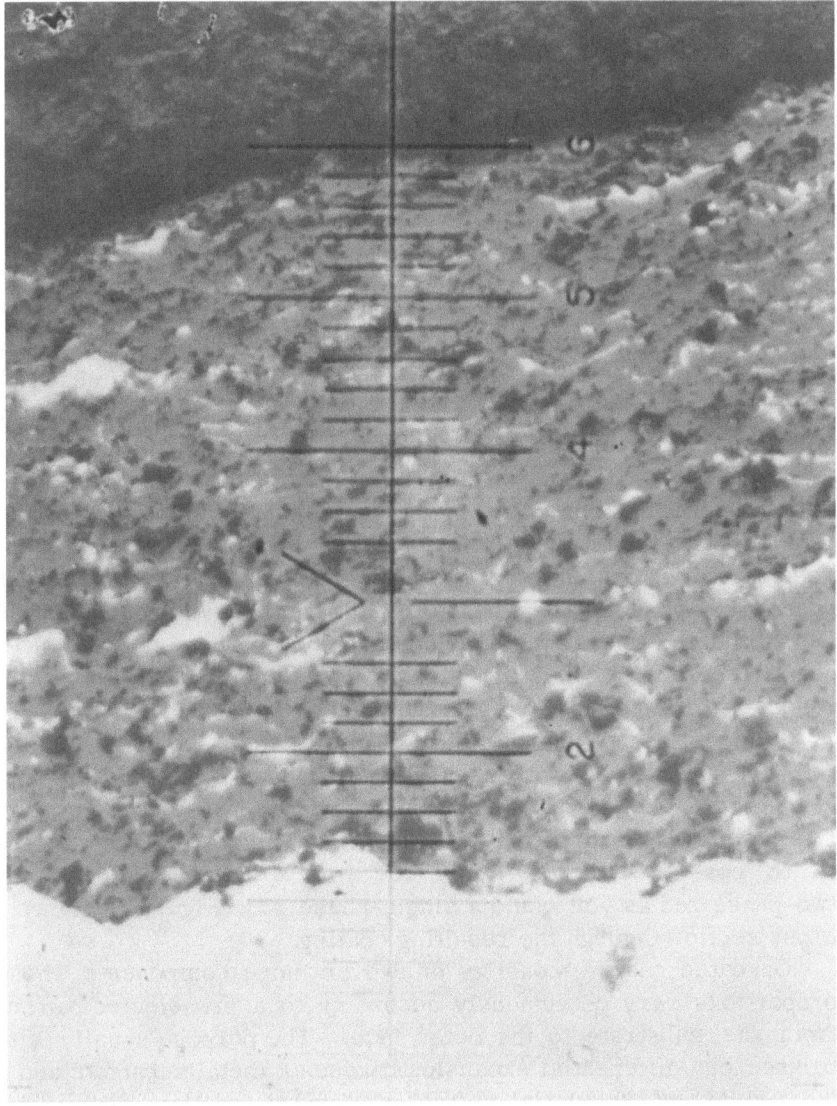

Fig. 14. Rod-sprayed zirconia—molybdenum coating cross section (1 division=30 μ).

Nuclear power systems for space vehicles appear well on the way, and these will require lightweight shielding easily attachable to structural parts.

Boron carbide is considered a useful shielding material since it contains the boron isotope B^{10}, an effective absorber of thermal neutrons.

Boron carbide can now be flame-sprayed by the flame-spray processes and appears ideally suited for such applications when its low density is also taken into consideration.

VII. MULTIPHASE COATINGS

In addition to the single-phase type of coatings covered thus far, the urgent need for aerospace materials with different characteristics is spurring much activity in the development of multiphase coatings. New concepts include laminated coatings, composite resin-metal-ceramic coatings, and gradated coatings.

Multilayer coatings of alternate layers of molybdenum and alumina appear to have combined the erosion resistance and excellent bonding properties of molybdenum with the good thermal shock resistance and insulating properties of aluminum oxide. One such coating composed of twelve layers with a total thickness of 0.050 in. has withstood rocket motor blasts at 4500°F in tests at Chance-Vought Aircraft, whereas single-phase coatings in the same thickness of either material fail immediately.

Composite coatings of metal–ceramic, resin–ceramic (plastic impregnation), etc. are frequently developed to change thermal conductivity, reduce porosity, and increase abrasion resistance. One unique way of creating a composite coating by the rod process is to coat a ceramic spray rod with a metal, then spray the two-phase rod as you would a single-phase rod. Figure 14 shows a cross section through the resulting coating.

Gradated coating consists of two or more components whose proportions vary continuously according to a preselected pattern from the substrate to the outer face. The purpose usually is to balance out differential expansion between a metal substrate and a desired thermal insulating refractory ceramic coating. Such a type of coating involving molybdenum–nickel chromium–zirconia has been successfully operated on the X15 power plant.

Thus, if we classify new materials as those which while known, are still undeveloped for space or commercial application, continued progress can be expected in this relatively new field of flame-sprayed ceramics.

High-Temperature Materials and Flame Spray Coatings for the Aerospace Industry — Their Processing, Characteristics, and Applications (Part II)

R. W. Love, C. C. Esty, and W. M. Wheildon

Norton Company, Refractories Division, Worcester, Massachusetts

Ceramic materials are now coming into their own in the aerospace industry. Solid ceramics in combustion chambers and throats are replacing metals because of their extremely high fusion or decomposition temperatures. Coatings supplement both ceramic and metals by adding such properties as thermal insulation, erosion resistance, chemical resistance, and, in the case of metals, a temperature-resistant property.

Data is given in Part I on oxide coatings such as alumina, zirconia, chromia, and nickel oxide. This pertains to characteristics important to the aerospace industry, such as optical refractoriness and mechanical, electrical, and neutron shielding. Coating systems are explained in relationship to basic concepts and methods. Advantages of newer methods of ceramic coating techniques and their relationship to the aerospace industry are brought forth.

Information is given in Part II on solid ceramics such as the carbides, borides, and oxides. This includes the relationship of manufacturing technique to such properties as thermal shock, erosion resistance, and strength. Other properties important to aerospace applications, such as oxidation resistance and melting point, are categorized. Recrystallized silicon carbide now presently used in aerospace systems is discussed.

I. REFRACTORIES

High-temperature materials and materials for aerospace use are terms that have become somewhat synonymous. Large sources of energy are needed to launch "hardware" into various required space trajectories. This means, first of all, that high temperatures will be encountered, at least until a cryogenic engine is developed, which does not seem likely in the near future. Means of propulsion are also needed to guide vehicles in space; this again signifies high temperatures. The third major contribution to heat in aerospace is the high temperatures produced upon entrance into the relatively high-density gas envelopes surrounding celestial bodies.

The highest-temperature materials known to us at the present time as a total class are a group of inorganic compounds and mixtures of compounds commonly called ceramics, or, to be slightly more accurate in this sense of use, refractories. These compounds include oxides, carbides, borides, nitrides, silicides, beryllides, and sulfides.

High-temperature materials are required in the following parts of an aerospace vehicle: (1) leading edges, (2) certain body structural components, (3) nose cones or re-entry type configurations, and (4) nozzles and other components in propulsion systems.

We have emphasized refractories as high-temperature materials. They do have other properties which are quite important in aerospace hardware.

1. Most of the species are hard and offer high resistance to wear by abrasion with solids as well as resistance to gas erosion.

2. As a class, these compounds are lighter than their respective metals, which is advantageous in cases where weight is a penalty, as it usually is.

3. Certain species are good thermal and electrical insulators, while other species are good thermal and electrical conductors.

4. Certain species have high oxidation resistance; others work well in a reducing environment.

5. Due to numerous recent technological advances, refractories can be fabricated in a number of different forms—fibers, foams or cellular structures, coating systems, and composites with metals or organic compounds or both.

There are, however, a number of major drawbacks to be considered.

1. As a class, refractories are brittle materials and therefore offer quite unique problems to the designer. High strength in compressive loading is common, but tensile properties are poor.

2. Refractories, particularly those of the high-temperature varieties such as the borides, carbides, and nitrides, are primarily manufactured by batch-processing techniques, and uniformity of the product still leaves much to be desired. This can be seen by the broad ranges of property data that can be found in the literature on these high-temperature compounds.

3. In reference to the borides, carbides, nitrides, or the materials which Schwarzkopf and Kieffer have labeled the "refractory hard metals," very little is understood of the effect of impurities and stoichiometry on properties. Most of the material that is available commercially has relatively high impurity levels, which cause grain boundary effects and can stimulate corrosion in these

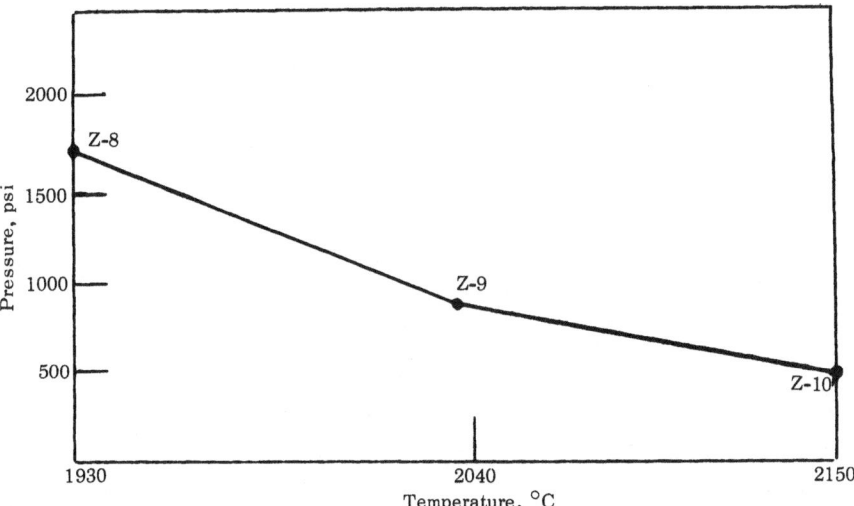

Fig. 1. Pressure-temperature curve for zirconium diboride, time—60 min. At 1680 psi, sample Z-8 had a density of 5.96 g/cc and a 8-μ crystal size. At 920 psi, sample Z-9 had a density of 5.95 g/cc and a 18-μ crystal size. At 610 psi, sample Z-10 had a density of 5.84 g/cc and a 25-μ crystal size.

areas. Also, much of the data that exists on these compounds has been gathered in the 1000—2000°C range, and yet we talk about using them at temperatures of 2200°C and up.

Figure 1 shows the effects of pressure and temperature at constant time on a hot pressing cycle for ZrB_2—the higher the pressure and the lower the temperature, the smaller the grain size. With a fall in pressure and a rise in temperature, a grain growth of approximately threefold is noted with, however, only a minor change in density. Figure 2 shows a photomicrograph of the structure of the ZrB_2. The fine 8-μ crystals are shown on the top left, the coarser 18-μ and 25-μ crystals at the bottom left and top right, respectively. Initial testing on thermal shock capabilities of 4-in. cross-section pieces using a liquid metal cycle test indicates that the thermal shock resistance of larger crystals far exceeds that of the finer. No spalling was noted on the 25-μ grain specimen after testing of a hundred cycles to 1200°C under plunge conditions and moving air cooling, whereas the finer structure began to shell after a very few cycles. Work is being continued to optimize boride microstructures in respect to strength and thermal cycling properties.

4. The aerospace designer is also invariably faced with problems, such as the need for a thermal insulator at a temperature

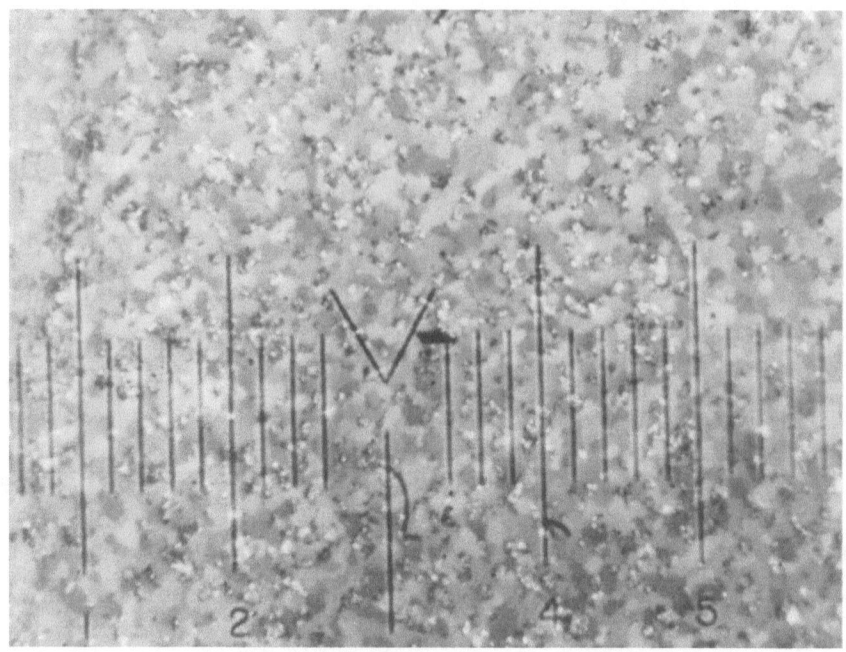

Fig. 2. Zirconium diboride photomicrograph.

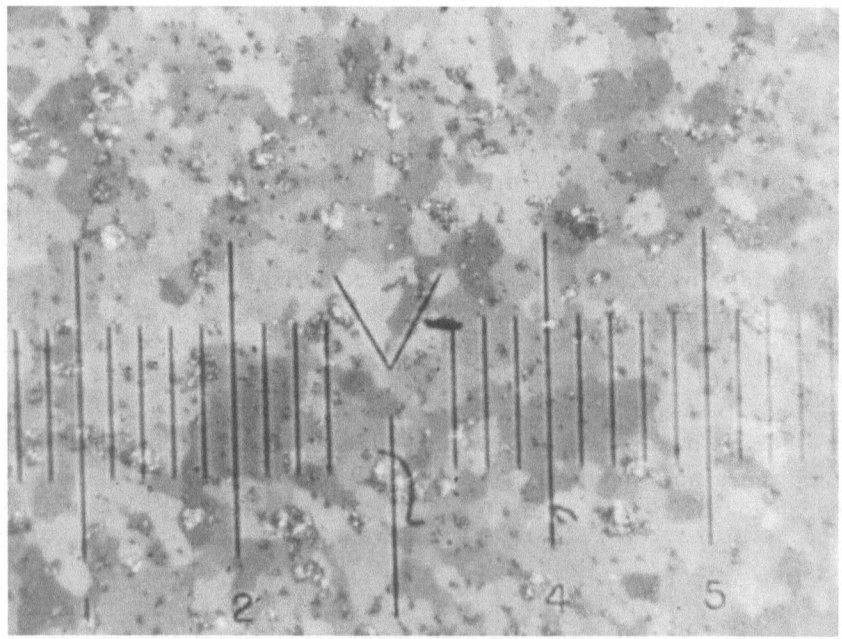

Fig. 2. Continued.

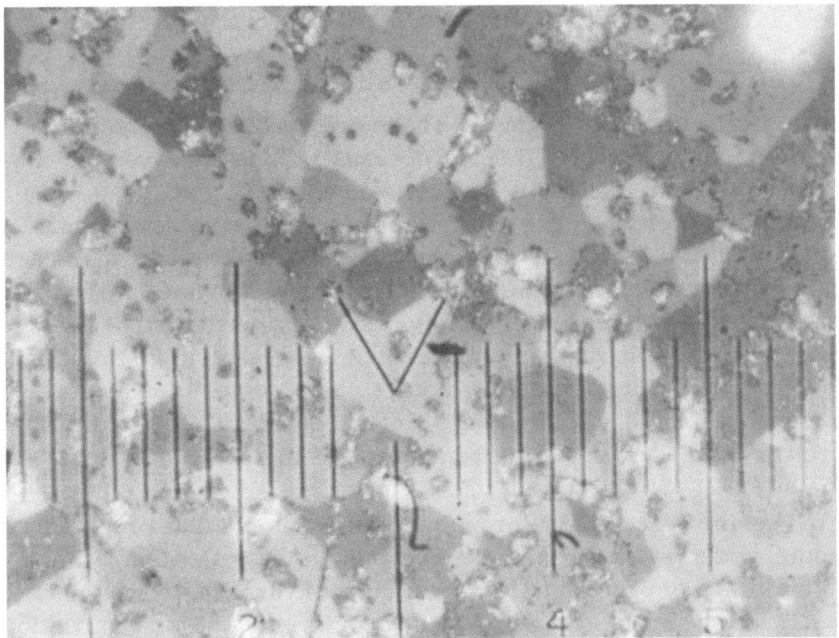

Fig. 2. Continued.

in excess of 2500°C in oxidizing atmospheres.   If one refers to tables of properties, one finds that nearly all the materials available are thermal conductors and all tend to oxidize fairly rapidly. Therefore, much more work is still necessary on such things as composite or refractory alloy structures where divergent properties of materials can be compromised to suit the needs of an application.

An interesting method of examining developments in the aerospace industry is to consider a specific case history and follow the steps that have been taken toward initial solution and the steps that are being planned to carry this solution further on.   In aerospace material work, there is no lasting solution to an application area.   Frequently, as fast as one finds the optimum material, someone else changes the requirements of the mission. New requirements to contend with may be new configurations, a completely different thermal envelope, or a new propellant that burns several hundred degrees hotter with a new chemical environment.   Each so-called solution starts a new chain reaction in the materials field.

II. SPACE ENGINE REQUIREMENTS AND MATERIAL SELECTION

One of the higher priority and most interesting fields in the nation's space program at the present time is that which is categorized as space engines. These are systems using mono- and bi-propellants for low thrusts necessary for space-vehicle maneuvering, course changes, and attitude control.

The problems in this type of application are somewhat unique. Simplicity and reliability are the key words. These systems are part of a vehicle's payload, so that weight is of prime importance. This makes complex fuel cooling systems unattractive. Space motors in a great number of cases must take a variable propellant flow for throttling purposes. Motors on space vehicles are turned off and restarted several times. This results in severe thermal cycling from subzero space temperatures to 2000°C and higher, in a matter of a few seconds.

Present space engine propellants are classified as prepackaged liquid systems—nitrogen oxide, hydrazine types. Gas temperatures in the range of 5100°F (2815°C) are noted; hydrogen, oxygen, nitrogen, carbon monoxide, and water vapor are the major combustion products.

A strict requirement of low thrust space engines is a high degree of reliability in the throat. For example, small 25-lb thrust nozzles will have throat diameters of approximately $\frac{1}{4}$ in., so that any small erosion whatsoever will make a major change in the thrust area and gas velocity. The material solution to this problem was not an easy one. Any metals that could withstand the environmental conditions were too heavy. Resin-impregnated inorganic fiber-type structure would not retain the throat diameter, due to material loss and flow in the ablating process. Graphite and carbon products closely approached the application needs, but there was enough throat burn-out and gas erosion to cause excessive throat washout, and performance was seriously affected. Certain oxide ceramics, such as zirconia and thoria, had the ability to withstand the temperature conditions and also the gas environment. However, due to their low thermal conductivity, thermal shock became a serious drawback. These oxides were not reliable in space engines where restarting was necessary, as they could not survive the thermal cycling.

This left the refractory hard metals group of borides, carbides, and nitrides. The material that was selected was silicon carbide

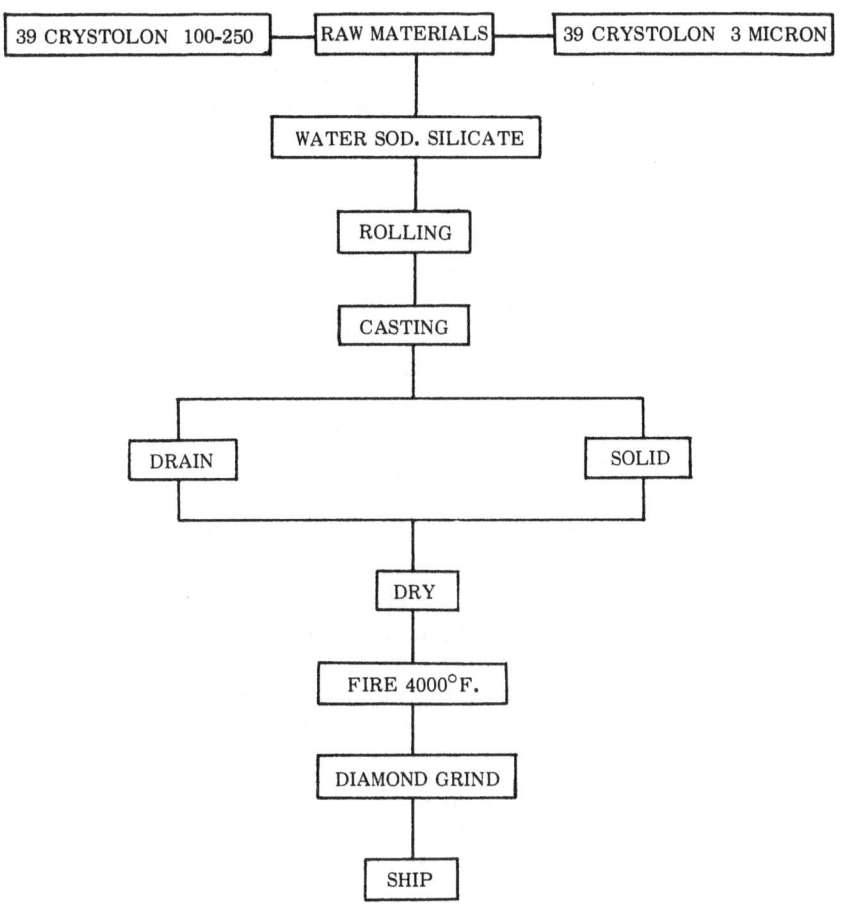

Fig. 3. Silicon carbide flow sheet.

because it had the following characteristics: (1) high thermal conductivity and resistance to thermal shock, (2) low thermal expansion, (3) high degree of oxidation resistance (at least in reference to other members of the carbide family), (4) long history of commercial, raw-material production and a known ability to make a uniform, high-purity silicon carbide grain product from which to fabricate shapes, and (5) lightweight material.

Most of the commercial silicon carbide products that were commonly used up to this time were of the bonded type, e.g., silicate or glass bonds and silicon nitride-type bonds. This was due to the fact that silicon carbide under standard conditions decomposes rather than forming a liquidus. This made self-bonding a very difficult feat using normal ceramic firing procedures.

However, neither of the previously mentioned materials could be considered for space engine work because neither bonding system could survive in the temperature range of the burning propellant.

During the middle 1950's the Norton Company developed a method to produce a recrystallized silicon carbide product requiring no bond. This process was further developed so that nozzle inserts for-space engines could be produced.

Figure 3 shows a flow chart for the production of silicon carbide throat inserts. A 100—250-grit green silicon carbide product is mixed with a fine material of approximately 3 μ sizing. These are dry-blended and then mixed with water and a minor amount of sodium silicate to form a slip from which parts can be cast in plaster-of-Paris molds. Depending upon the configuration desired, either drain or solid casting techniques can be used. The parts are dried to remove all moisture and then fired to temperatures exceeding 4000°F (2205°C) in an induction–type furnace under special atmospheric conditions. This allows a recrystallization and bonding of the silicon carbide to give a dense, fine-pore-structure body. Where necessary, close tolerances and special shaping are done by diamond grinding.

Table I lists the major properties of recrystallized silicon carbide, which is now being used on a number of space engine systems with success. Thermal shock failures are low, and the erosion rate under chamber pressures of 100 psi is low.

As previously stated, there really is no final material solution in the aerospace business. Designers are now asking for greater chamber pressures; and, instead of present cumulative running times of 300 sec, they want periods in excess of 1000 sec. The recrystallized silicon carbide throat could not maintain itself under these conditions, so we now start to examine composite systems.

One example of a composite system is a silicon carbide throat impregnated with WSi_2. This particular composite structure has operated for long periods of time in excess of those attained with recrystallized silicon carbide at space engine chamber pressures. The WSi_2 fills the fine SiC pore structure to give a dense, strong body with high erosion and oxidation resistance.

Where do we go from here? The chemical engineers now tell

TABLE I

Properties of Recrystallized Silicon Carbide

Property	Description
Maximum Operating Temperature	4400°F, 2450°C (reducing conditions) 3200°F, 1750°C (oxidizing conditions)*
Bulk Density	2.4 – 2.6 g/cc, 155 to 168 lb/ft^2
Permeability	Slight (can be controlled)
Porosity (total)	18 – 25%
Surface Finish	Few hundred microinches without grinding
Abrasion Resistance	Very high
Modulus of Rupture	17,000 – 20,000 psi, 20 – 2250°C
Modulus of Elasticity	23 x 10^6 psi at room temperature
Compressive Strength	75,000 – 100,000 psi at room temperature
Hardness	2500 (Knoop at 100 g)
Thermal Conductivity	115 Btu/hr/ft^2 - in./°F
Linear Thermal Expansivity	2.8 x 10^{-6}/°F Mean 20 – 2700°F
Thermal Shock Resistance	Excellent
Electrical Resistivity	Approximately 3 ohm-cm at 70°F Increases with temperature
Chemical Resistance	High resistance to acids and oxidation. Some oxidation will occur at prolonged high temperature.

*Crystolon "R" silicon carbide has operated at temperatures up to 3200°F with oxidizing flames with only minor attack.

us that they have new propellants on the way. Rather than nitrogen oxide-type oxidizers, we now face oxygen difluorides which will mean much higher propellant burning temperatures (6500°F, for example) as well as combustion products such as fluorine and hydrogen fluoride and, in the difluoride boron hydride system, a number of boron and fluorine radicals.

One possible approach to this might be to attempt the same type of fabrication philosophy in a much higher fusion point material, such as TaC and HfC. However, it is questionable whether we will be able to protect these from severe oxidation for the lengths of time required with any type of coating system that could survive.

Another approach is to retain our present hard silicon throats, but to form them in a body of high-porosity, recrystallized silicon carbide with a continuous pore structure which can be filled with an ablative-type compound. Pore diameter controls the ablative cooling system. Ablatives streaming from this system could keep the denser hard throat insert cool enough to survive in the environment of new propellants for space engines.

Boron Nitride for Aerospace Applications

J. E. Fredrickson and W. H. Redanz

Carbon Products Division, Union Carbide Corporation
New York, New York

This paper describes the properties of new grades of boron nitride now available in large sizes in production quantities. The properties of hot-pressed boron nitride made to a controlled chemical composition in large sizes are described. Pyrolytic forms are also included in the discussion. Emphasis is on new high-temperature property data including coefficient of thermal expansion, a high-temperature permanent set, thermal conductance, and thermal diffusivity. New helium permeability data are also presented. Experimental efforts on metallizing and brazing to the surface are discussed, and also methods to avoid spalling due to moisture pickup. Applications include plasma arc insulators, radar windows, heat sinks, and thermoelectric insulators.

I. INTRODUCTION

Boron nitride is the only available engineering material that is readily machinable, nontoxic, a good conductor of heat, and an excellent electrical insulator. It not only meets many demanding high-temperature engineering needs, but also finds numerous low-temperature uses as well.

New grades of hot-pressed boron nitride, available in the last eighteen months from Union Carbide Corporation's Carbon Products Division, have further enhanced the potential usefulness of this material. One grade in particular has already gained wide acceptance for its superior performance when used for insulators, spacers, and containing channels in plasma arc heaters. Another new grade demonstrates outstanding resistance to moisture pickup (which can cause spalling) and improved high-temperature properties. Sufficient physical data, including information on the phenomenon of permanent thermal expansion, are now available for the designer to consider boron nitride for many new uses.

The performance of the new grades of material as plasma arc insulators suggests their evaluation for high-temperature thermal-shock applications, such as heat sinks and thermal radiation shields. The outstanding dielectric properties which have proved useful in

245

low-temperature electronic applications also offer promise in applications for a high-temperature antenna material.

II. BACKGROUND

Although boron nitride compounds were studied at the turn of the century, these materials did not appear in research quantities until the mid-1950's. The primary emphasis of the work at this time was to develop coatings for high-temperature electrical insulators and metallurgical crucible linings. This early work showed that boron nitride was indeed among the best of the high-temperature insulators, but that the available techniques of making boron nitride were not economically reliable.

In the late 1950's, several laboratories concentrated on hot-pressing techniques. Solid molded shapes in limited sizes and quantities soon became available. It was these shapes which gave the application engineer the opportunity to test this new material. The main result of this experience was to demonstrate hot-pressed boron nitride as a useful engineering material for high-temperature use. In addition, the material's unusually low loss tangent and low dielectric constant in combination with high thermal conductivity proved useful for electronic applications.

As is common with most new materials in their early stages, practical applications were hampered by the usual problems arising from lack of experience. For example, one of the early problems that arose was the effect of moisture pickup on high-temperature performance. In addition, larger stock was required in many cases.

In the spring of 1963, "National" Grade HBN became available in the largest size yet produced—14 in. in diameter and 12 in. long. This is the grade of boron nitride that has gained wide acceptance for premium performance in high-temperature applications. It is available from stock in a variety of rod, rectangle, and plate sizes.

III. BORON NITRIDE

Boron nitride is a white chemical compound that is fairly inert and stable. It forms a six-membered ring of alternate boron and nitrogen atoms and has a hexagonal crystal structure.

The powder can be prepared in many ways. Three potentially commercial methods are illustrated by the following equations:

$$B_2O_3 + 2NH_3 \xrightarrow[\text{[Ca}_3\text{(PO}_4\text{)}_2 \text{ filler]}]{(800 - 1200\,^\circ C)} 2BN + 3H_2O \qquad (1)$$

$$BCl_3 + 4NH_3 \xrightarrow{(1600 - 1900\,^\circ C)} BN + 3NH_4Cl \qquad (2)$$

$$B_2O_3 + Ca(NC)_2 \xrightarrow{(800 - 1500\,^\circ C)} 2BN + CaO + 2CO \qquad (3)$$

The melting point of boron nitride powder is reported to be in excess of 3000°C under elevated pressures of nitrogen. At 1 atm of nitrogen, it is reported to sublime at 3000°C. Boron nitride powder itself has found little commercial usage. It is in the compacted and sintered form that it has found application as an engineering material.

IV. HOT-PRESSED BORON NITRIDE

Hot-pressed boron nitride is compacted at temperatures up to 2000°C and at a 1000-psi pressure [1]. In addition to the conditions of pressure and temperature, the presence of oxygen is required to form a dense strong engineering material. This effect can be described in terms of mixtures of boron nitride powder and boric oxide (Table I).

At present, 100% pure hot-pressed boron nitride shapes are not available.* All of the marketed hot-pressed shapes are actually composite materials. This is a very important point. Published literature on the properties of boron nitride shapes to date generally does not provide a complete chemical analysis for the pieces being studied and, at times, does not include the method of manufacture. One reason for this has been the difficult problems encountered in the chemical analysis of boron nitride systems.

"National" grade HBN hot-pressed boron nitride is produced to a specified minimum density of 1.90 g/cc and the following chemical composition (expressed as wt.%): boron—42.4–44.4, nitrogen—53.3–54.7, oxygen—3.9 (max.), carbon—0.4 (max.), and impurities—

---

*It should be pointed out, however, that pure boron nitride shapes are available in the pyrolytic form. "Boralloy" pyrolytic boron nitride is an example of a commercially available form. This material is made at a temperature of 3542°F and a pressure of 1 mm Hg. It is formed on graphite surfaces held at this temperature by the chemical reaction of boron trichloride and ammonia. The deposited material is polycrystalline, pure (100 ppm impurities), hexagonal boron nitride with high crystal orientation and anisotropic properties.

TABLE I

Effect of Boric Oxide on Hot-Pressed
Boron Nitride Powder [1]

BN + % B_2O_3 added	Density (g/cc)	Weight loss (20 hr) 70°C in methanol (%)
0.0	1.22	nil
1.0	1.58	-
2.5	1.77	nil
5.0	2.07	2.95
10	2.08	6.2
15	2.06	9.4

0.05 (max.). This is the grade to be discussed in detail. It is hot-pressed in a production plant operation up to 14 in. in diameter and 12 in. long. A second grade, HBR, with similar chemical properties, is available in production quantities and will be discussed with respect to improved moisture resistance and high-temperature properties.

V. MECHANICAL PROPERTIES

"National" grade HBN hot-pressed boron nitride can be considered a fine-grained material. It is very much like fine-grained graphite, except that room-temperature strength moduli are two to three times greater than the best graphites. (This is not true at higher temperatures as will be noted later.) It is a material with a hardness in the range of gypsum, and it can be machined easily with ordinary steel cutting tools.

Table II summarizes the mechanical properties of this material. Note that the degree of anisotropy is low. Grade HBN is more isotropic than any other boron nitride previously reported in the literature. As is characteristic of this type of material, careful sample preparation is required for reproducible results when making these measurements. Sample shapes having sharp changes of contour should be avoided, and the surface finish should be reproducible. These same factors should be taken into consideration in application design, if maximum material strengths are to be utilized.

Flexural strength measurements were obtained on a Tinius Olsen strength testing machine fitted with center point loading ap-

TABLE II

Typical Mechanical Properties Grade HBN Boron Nitride
(Room Temperature)

A is the force applied perpendicular to hot-pressing direction.
B is the force applied parallel to hot-pressing direction.

Property	Units	Value	Ratio A/B	Notes
Bulk density	g/cc	2.0		
Flexural	psi	$16,000^A$ $14,000^B$	1.14	t s/t = 3
Tensile	psi	$6,500^A$ $8,000^B$	0.82	I I/d = 1.2
Compressive	psi	$27,500^A$ $26,500^B$	1.04	I/d = 2
Young's modulus	10^6 psi	11^A 9^B	1.21	Sonic method

paratus. Blunted knife edges ($\frac{1}{16}$-in. radius replaceable edge) were used with a 1.5-in. span. Samples tested were 0.500-in. by 0.500-in. cross section, giving an approximate span-to-sample-thickness ratio of 3.

Compressive strength was also measured with the same machine equipped with a pair of plattens. The lower platten was provided with a ball joint to allow for slight end variations in parallel alignment of the samples. The boron nitride samples were cylindrical in shape, 0.500 in. in diameter and approximately 1 in. long. Previous testing on graphitic materials has indicated that varying the length-to-diameter ratio from approximately 1 to 2 causes no significant change in compressive strength results.

Tensile strength testing results can be greatly influenced by specimen geometry. The specimen configuration selected for the testing of boron nitride is that presently under consideration by ASTM Committee C5, Subcommittee 4, as a recommended specification for fine-grain mechanical carbon and graphite.

VI. STRENGTH PROPERTIES VS. TEMPERATURE

Figure 1 illustrates the change of flexural strength with temperatures up to 2000°C. Grade HBN shows a rapid drop in flexural

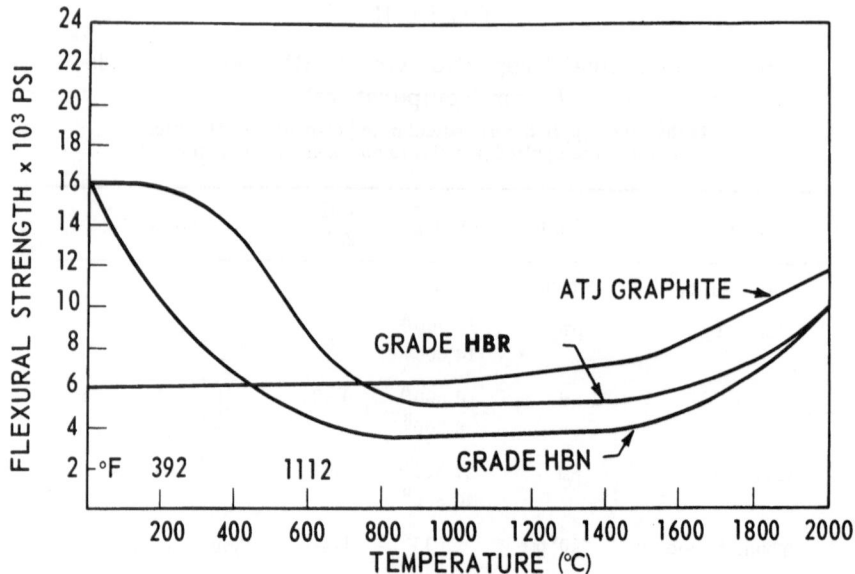

Fig. 1. Flexural strength vs. temperature.

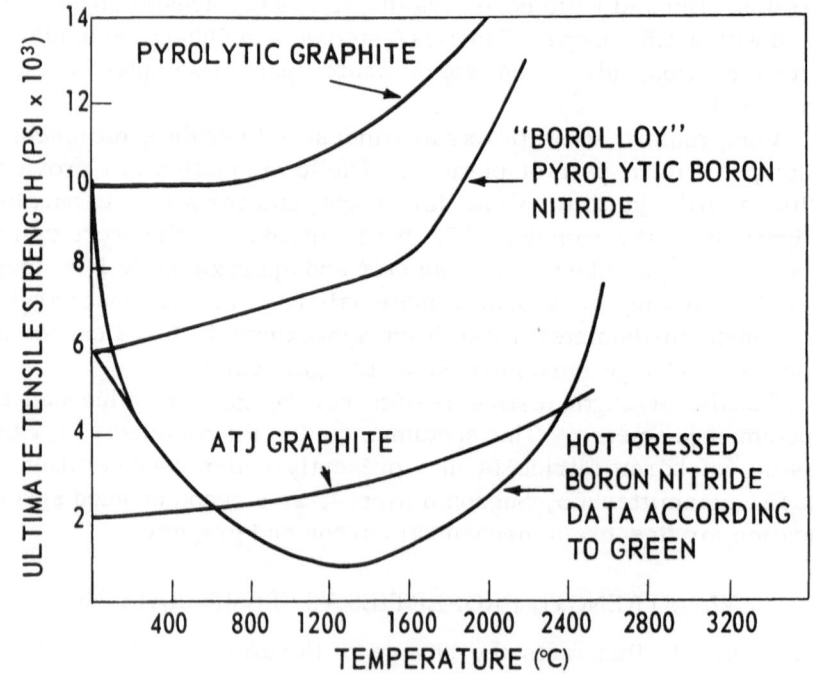

Fig. 2. Ultimate tensile strength vs. temperature.

strength from room temperature to 600°C (1112°F). At this point, the flexural strength then remains relatively constant until 1600°C (2900°F). Above 1600°C, the flexural strength increases until, at 2000°C, it is more than double the lowest strength.

The new grade HBR material shows higher strengths up to 200°C (392°F) and then drops at 600°C (1112°F). It also levels off in the 600—1600°C range, although at a higher level than grade HBN. At temperatures above 1600°C, the two grades have the same higher strengths. This strength improvement at higher temperatures for boron nitride shapes is very similar to graphite, as illustrated by the curve for ATJ graphite.

The increase in strength above 1600°C has also been noted in the tensile strength studies of Green [2] for hot-pressed boron nitride and Kotlensky [3] for pure pyrolytic boron nitride (Fig. 2). Green's data also indicate that the influence of grain orientation disappears at high temperatures for hot-pressed boron nitride.

VII. COEFFICIENT OF THERMAL EXPANSION AND PERMANENT SET

In designing for high-temperature application, attention should be given to thermal expansion characteristics. When hot-pressed boron nitride is heated for the first time above 1200°C, a portion of the thermal expansion remains as a permanent expansion. This is illustrated in Fig. 3 for a sample of grade HBN heated to 1800°C.

Note that, of a typical total expansion of 1.5% at 1800°C for grade HBN, 80% remains as a permanent set at room temperature. When the same sample is run again, this permanent expansion is about 25% of the total 1800°C expansion, and only 10% of the first permanent expansion.

The thermal expansion characteristics are illustrated in more detail in Figs. 4 and 5. These curves plot the percent expansion versus temperature perpendicular to, and parallel to, the pressing direction. The slope of the resulting plot is the coefficient of thermal expansion. Note the sharp changes in slope in the 1100—1200°C degree range. This slope change occurs at 1500—1600°C in a sample preheated to 1800°C.

In the sample preheated to 1800°C, the coefficient of thermal expansion is reduced 50% in the room-temperature to 1600°C range, and approximately 33% in the 1600—2000°C range. These values are probably more representative of the true reversible coefficient of

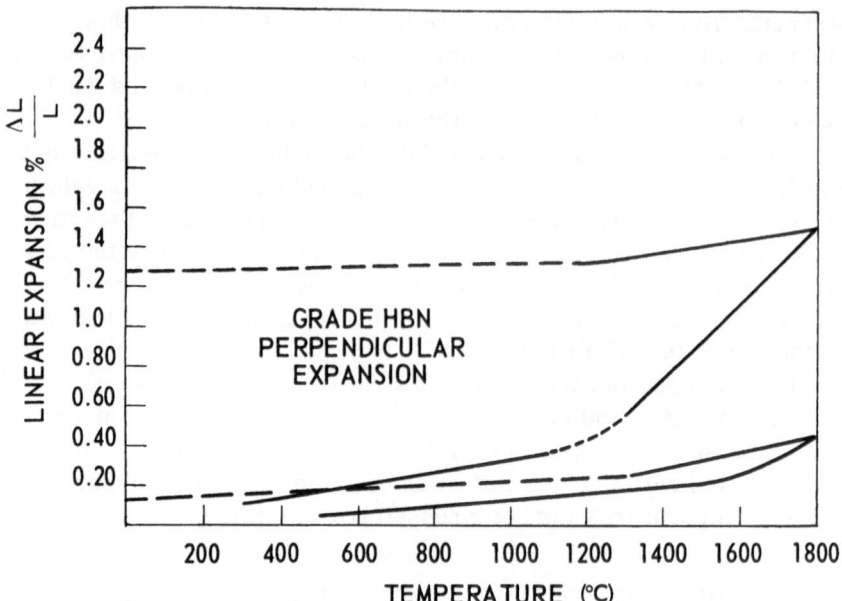

Fig. 3. Typical percent of thermal expansion data plotted versus temperature.

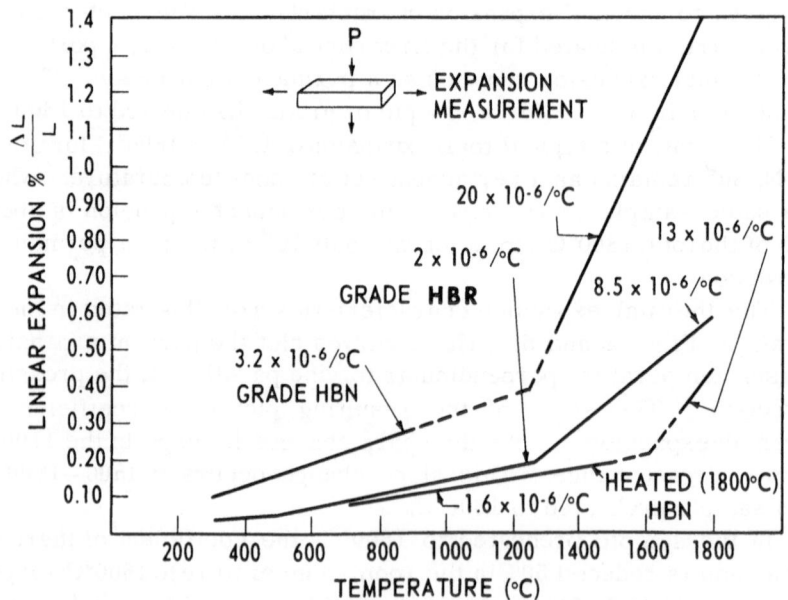

Fig. 4. Typical thermal expansion characteristics of "National" hot-pressed boron nitride.

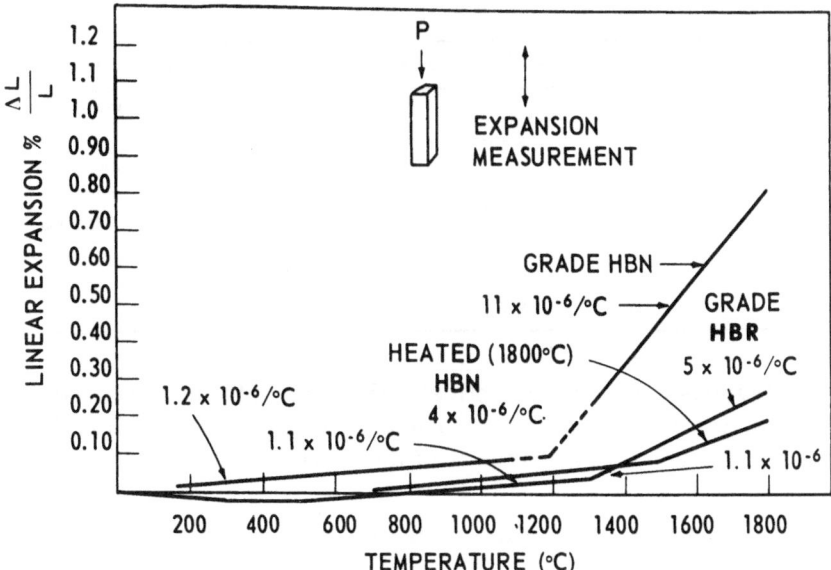

Fig. 5. Typical thermal expansion characteristics of "National" hot-pressed boron nitride.

thermal expansion. Grade HBR has approximately the same C.T.E. up to 1200°C as the preheated sample. It also has a lower C.T.E. than grade HBN or the heated sample at high temperatures.

Table III summarizes the data in Figs. 4 and 5 and compares the expansion characteristics in opposite directions for both grades. At high temperatures, the C.T.E. of grade HBR is approximately 45% that of grade HBN in both directions. The permanent set is approximately 25% of that for grade HBN. These lower values should result in improved thermal shock resistance at high temperatures.

Also, note that the comparison of values in both directions shows that the C.T.E. values for grade HBN in the perpendicular direction are 2 to 3 times that in the parallel direction. This difference is much lower than factors of up to 20 reported by Ingles and Popper [1]. A similar low degree of anisotropy for grade HBN was also shown in the mechanical property data.

The method employed for obtaining the expansion data is widely used for high-temperature measurement of graphites, ceramics, refractories, and metals. It is presently under consideration as a standard test method by ASTM Committee C5, Subcommittee 3. In this testing method. the sample rests on knife edges in a small resistance-heated graphite tube furnace. Sample length measure-

TABLE III

"National" grade	Temperature range (°C)	Perpendicular Expansion		Parallel Expansion	
		First heatup 10^6 °C^{-1}	Permanent set 1800°C − R.T.	First heatup 10^6 °C^{-1}	Permanent set 1800°C − R.T.
HBN	0−1200	3.2 ⎫		1.2 ⎫	
HBN	1200−2000	20 ⎭	1.25%	11 ⎭	0.75%
HBR	0−1200	2 ⎫		1.0 ⎫	
HBR	1200−2000	8.5 ⎭	0.30%	5 ⎭	0.20%

ment changes are obtained optically through sight tubes inserted at right angles to the sample. Measurements are taken on both the heating and the cooling portions of the testing cycle.

VIII. MOISTURE PICKUP AND SPALLING

In the 1958−1960 period, evaluation tests on the thermal shock characteristics of hot-pressed boron nitride were often in conflict. Reports ranged from poor thermal shock resistance at 600°C [1] to excellent resistance at 2000°C. The reports generally describe a condition in which surface pieces of boron nitride literally blew out, leaving pockmarks or pits.

Moisture absorption was early suspected as the cause. Some confusion did exist because 100°C-drying did not always seem to eliminate the problem. However, it was found that heating parts to 350°C eliminated spalling. For this reason, it is believed that the basic absorption phenomenon involves hydration of the boron oxide content by atmospheric moisture, so that the "drying" is really a dehydration reaction that requires a specific temperature at 1 atm.

Spalling becomes a problem only if very rapid heating of the BN is required. Apparently the moisture does not escape fast enough from beneath the surface, and the resulting pressure buildup ruptures surface layers, leaving small pits. In prolonged exposure to humid atmospheres, deeper penetration can result in larger pits and serious cracking.

Figure 6 illustrates the moisture pickup characteristics of grade HBN and grade HBR. In this test, accelerated moisture pickup was simulated by using a "wet box" at 85% relative humidity and room temperature. Samples measuring 2 in. by $\frac{1}{2}$ in. by $\frac{1}{4}$ in. were dried

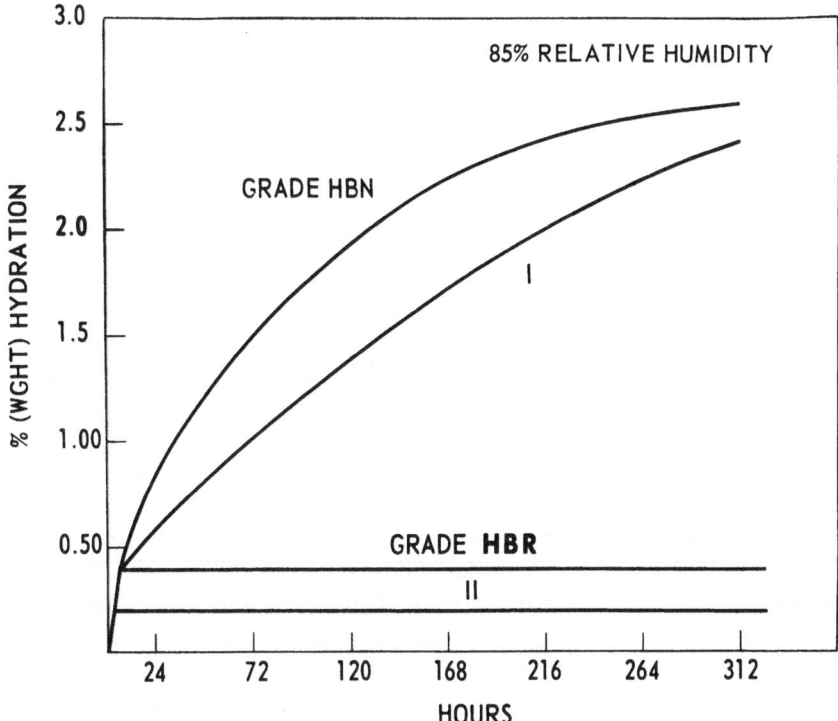

Fig. 6. Eighty-five percent relative humidity room temperature hydration.

at 350°C and weighed for "dry" sample zero points. The samples were then placed in the wet box and weighed periodically without disturbing the atmosphere.

Note that all samples picked up 0.3–0.4 wt.% moisture in the first two hours. Grade HBN then continued to pick up weight over a period of 14 days at decreasing rates. The tests were stopped at 14 days, at which time a 2.6 wt.% increase of grade HBN was recorded. If it is assumed that all the oxygen in this grade is present as B_2O_3, a maximum water pickup of 3.3% would be expected for a typical oxygen content of 3.0%.

Grade HBR is a significant step in the direction of eliminating moisture pickup. The total moisture pickup has been reduced to 0.4%—all of which occurs in the first four hours of exposure. This is believed to be simple surface adsorption in equilibrium with the vapor pressure. Curves I and II in Fig. 6 illustrate the possibility of reducing moisture pickup by protective chemical coatings.

A word of caution is necessary in interpreting the number shown in Fig. 6. The weight percent pickup refers only to the size sample used and the humidity. Larger samples with a lower surface-to-mass ratio would have a lower weight percent water pickup in a given period. This test is useful, then, for a comparison of the performance of different samples of equal size, and the data should not be used in an absolute sense.

A thermal shock test for moisture-laden parts was simulated as follows: A grade HBN sample was removed from the box after 7 days with a water pickup of 2.0%. This sample was lowered into a preheated tube furnace at 900°C. The sample cracked severely in 10 sec. For comparison, a sample of grade HBR removed after 7 days (0.4%) was similarly treated. It showed no change after 15 min in the furnace.

It is also interesting to note that grade HBR has a lower density (1.85) than grade HBN (2.0). This is a further indication supporting the belief that moisture spalling is mainly controlled by the hydration of B_2O_3 and is less dependent on the mechanical factors of surface finish and density. Grade HBR parts are now in field test as plasma arc insulators. Initial reports are highly encouraging, with users reporting superior performance where moisture pickup has been present.

IX. THERMAL CONDUCTIVITY

Thermal conductivity of boron nitride is similar to stainless steel. There is considerable confusion in the literature as to the accurate values for thermal conductivity. An average of available data indicates that the thermal conductivity decreases from 20 Btu-ft/hr-ft^2-deg F at room temperature to 12 Btu-ft/hr-ft^2-deg F at 800°C. The conductivity levels off at this value up to 1500°C.

X. DIELECTRIC PROPERTIES

An outstanding feature of both grade HBN and grade HBR is the low dielectric constant and low loss tangent (tangent delta) they possess. Figures 7 and 8 illustrate typical values over a temperature range of 20—500°C in the 8.5-kMC range. The data once again indicate that grade HBN hot-pressed boron nitride is not as anisotropic as has been reported for hot-pressed items in general. There is only a 2% difference in both directions, and no change in the loss tangent.

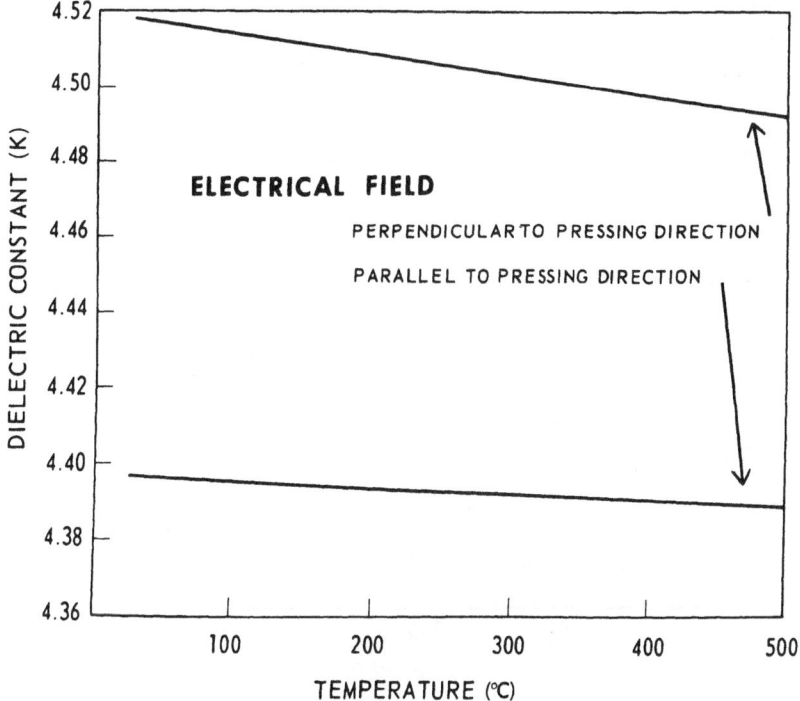

Fig. 7. Typical dielectric constants for "National" boron nitrides.

XI. ELECTRICAL RESISTIVITY

The electrical resistivity of boron nitride ranges from more than 10^{13} ohm-cm at room temperature to 10^2 ohm-cm at 2000°C. Reports from plasma arc users verify this desirable property, as well as a remarkable absence of tracking.

XII. HIGH-TEMPERATURE STABILITY

Little data has been published on the stability of hot-pressed boron nitride at high temperatures. Powers [4] reports that samples heated in air show little dimensional change for periods of 10—15 min in the 3000—4100°F range (1650—2260°C). In the 4100—4800°F range, he reports severe oxidation in 3—6 min periods.

Bro and Steinberg [5] report data on boron nitride in high-velocity atmospheric gases at enthalpies of 3800—9000 Btu/lb. Erosion rates of only 0.03 lb/sec-ft² in nitrogen, 0.06 lb/sec-ft² in synthetic

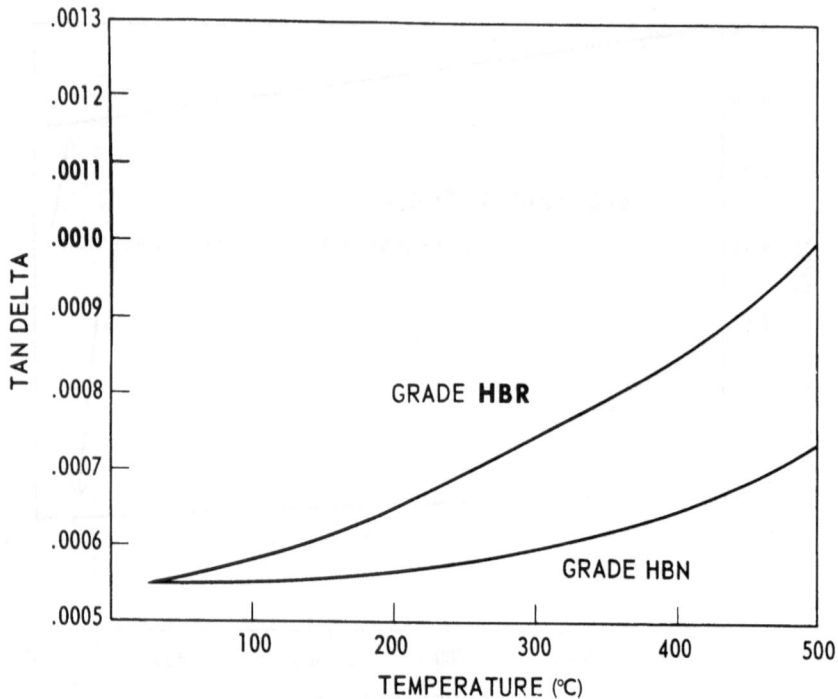

Fig. 8. Typical loss tangents for "National" boron nitrides.

rocket exhaust, and 0.12 lb/sec-ft^2 in air are reported at enthalpy contents of 8000 Btu/lb.

Analysis of the data indicates that 75% of the weight loss was due to spalling. That this spalling could be eliminated is indicated by Carter's report [6] that predried boron nitride walls undergo negligible material loss when exposed to nitrogen gas velocities of Mach 2 and a 6000°K stagnation temperature. In this work, the boron nitride walls form channels in a crossed-field plasma accelerator. Similar experiences are reported with the use of boron nitride tubes in plasma arc air heaters.

Any study of the stability of boron nitride at high temperatures should consider the following factors:

1. The grade and chemical analysis of the starting material.
2. The probable loss of B_2O_3 content in the hot-pressed material above 1300°C.
3. The glazing and possible protecting effect of B_2O_3 below 1300°C.

4. The decomposition rate of boron nitride itself. Schissel and Williams [7] show that the partial pressure of nitrogen is in the 10^{-4} —10^{-5} mm Hg range at 1700°C. The data extrapolate to $5 \cdot 10^{-4}$ at 2200°C.

5. The influence of a chemically-reactive environment such that air would be superimposed on these processes.

Bowen's recent theoretical study [8] of the combustion of pyrolytic boron nitride indicates that the decomposition reaction may serve to induce a large increase in the oxidation rate in the 2600—2800°C range. This may account for Power's observation [4] of the serious degradation of hot-pressed boron nitride in the 2200°C range.

In this respect, we can add our own experience with a composite material containing boron nitride. It is commercially used at 1600—1800°C as a resistance heater in a vacuum of 1 μ. No effect is noted on the gas pumping load. The heaters are used for hot periods of 12—24 hr.

Basche and Schiff [9] report oxidation rates of 0.1—1.0 mg/cm^2-min in the 1400—1600°C range for pyrolytic boron nitride compared to a value of 10 for pyrolytic graphite. The rate ranges from 1 — 10 mg/cm^2-min in the 1600—2000°C range. They also report that the presence of 1% carbon reduces the rate by 33% at 2000°C.

Again this points to the need for careful identification of hot-pressed boron nitride by chemical analysis for any high-temperature application tests. Grade HBN, for example, typically contains 0.2% carbon.

XIII. SOME BORON NITRIDE APPLICATIONS

Typical applications that have proved the usefulness of the high-temperature properties of "National" boron nitride include electrical insulators and wall liners in plasma arc devices such as gas heaters, arc jet thrusters, and high-temperature MIID devices [6,10—12]. As a wall liner, boron nitride allows higher wall temperatures for increased thermal efficiency. Its good thermal conductivity permits an improved response in controlling wall temperatures.

The very rapid buildup of temperatures in these plasma device applications demonstrates the outstanding thermal shock resistance of boron nitride. The gases involved in the devices are nitrogen, air hydrogen, and helium at subsonic to Mach 2 velocities. Because of the very slow erosion rate of boron nitride at high temperatures, reliable and predictable part-lives can be attained.

In resistor jet devices, boron nitride is used as the electrical insulating spacer for the tungsten resistance heaters. Large blocks of boron nitride are used and are reported to be operated in the 2200—2400°C range. Boron nitride spacers are also used to prevent stray arcing between the electrodes in three-phase AC arc heaters.

In the electronics fields, dielectric insulating cylinders of boron nitride are used around ferrite rods in Faraday gyrator devices. The high thermal conductivity of boron nitride allows the rods to operate at higher power levels without developing destructively high temperatures in the ferrite cores. The low dielectric constant and low loss tangent of boron nitride allow this material to be utilized as the dielectric medium in radar frequency antennas.

All of these applications have either direct or suggested thermal, chemical, and electrical uses in the aerospace field where boron nitride offers many advantages—light weight, corrosion resistance and high-temperature resistance, ease of machining to close tolerances, excellent thermal shock resistance to high heat flux, good thermal conductivity, low dielectric constant and low loss tangent, excellent electrical resistance at low and high temperatures, and production plant availability.

REFERENCES

1. T. A. Ingles and P. Popper, Special Ceramics, Academic Press, New York, 1960, pp. 144-167.
2. L. Green, Proc. Fourth Carbon Conf., Pergamon Press, London, 1960, p. 497.
3. W. V. Kotlensky and H. E. Martens, Nature, 196: 1090-1091 (1962).
4. D. J. Powers, Bell Aerosystems Co. Report 64-14(M), Oct., 1962.
5. P. Bro and Steinberg, J. Am. Rocket Soc. 31: 1460-1462 (1962).
6. A. F. Carter and A. P. Sabol, J. Am. Rocket Soc. 32: 424 (1962).
7. P. Schissel and W. Williams, Bull. Am. Phys. Soc. 4:139 (1959).
8. M. D. Bowen and C. W. Gorton, Combustion of Pyrolytic Boron Nitride, A.I.A.A. Conference, Preprint No. 63-486, Dec. 11, 1963.
9. M. Basche and D. Schiff, "New Pyrolytic Boron Nitride," Mater. Design Eng. 59(2): 78-81 (1964).
10. D. B. Langmuir, Electric Space Craft 1962, Astronautics 7:23 (1962); see also M. I. Yarymovych, F. A. deWiess, and R. R. John, Feasibility of Arcjet Propelled Spacecraft, Astronautics 7:36 (1962).
11. M. L. Yaffee, Electrothermal Thruster Rockets Reach Advanced Status, Aviation Week & Space Technology, 80 (6): 55 (1964).
12. C. E. Shepard, V. R. Watson, and H. A. Stine, Evaluation of a Constricted-Arc Supersonic Jet, NASA TND-2066, Jan., 1964.

Graphite-Base Refractory Composites for Aerospace Applications

D. C. Hiler and K. J. Zeitsch

Advanced Materials Laboratory, Union Carbide Corporation
Lawrenceburg, Tennessee

Graphite is one of the most widely used high-temperature materials because of its combination of refractoriness, thermal shock resistance, and machinability. Present-day applications, nevertheless, call for improvements of graphite with respect to oxidation resistance, strength, and hardness. Improvements of metals with respect to these properties have been accomplished by alloying; and, in a similar manner, the purpose of this study has been to enhance the properties of graphite by the use of additives. The manufacturing process employed was hot-pressing with final temperatures ranging from 1800–3200°C depending on the additives used, and pressures up to 10,000 lb/in were applied in some cases. The starting ingredients were powdery blends of graphite, pitch, and selected refractory materials not exceeding 30 vol. %. The most effective additives were zirconium, silicon, niobium, thorium, and some of the rare earth metals. The relative evaluation of the resulting composites was primarily based on oxidation studies. The latter were conducted in various apparatuses comprising resistance heating, arc image techniques, induction heating coupled with levitation and arc plasma exposure. The composites form an oxidation-resistant coating of the oxides of the additives which enables the material to withstand temperatures to 4000°F in an oxidizing atmosphere. Some of the composites have been applied as thrust chamber liners for liquid-fueled, attitude-control motors with success. The theoretical flame temperature of the fuels ranges from 4000–5300°F.

I. INTRODUCTION

Graphite is one of the most widely used high-temperature materials because of its unrivalled combination of refractoriness, thermal shock resistance, and machinability. Nevertheless, modern applications, especially in the aerospace field, call for improvements of graphite with respect to oxidation resistance, strength, and hardness. For metals, developments in this direction have been accomplished by alloying. Similarly, it has been the purpose of recent studies to enhance the properties of graphite by the use of additives.

According to such an objective, novel materials were prepared for which the carbon content was chosen generally to exceed 70 vol. % in order to preserve the basic features of graphite, especially thermal shock resistance and machinability. This amount of carbon is considerably more than can be dissolved by any additives. Therefore, the resulting products to be discussed here are composites rather than alloys, although extensive regional alloying was promoted by the application of processing conditions at which molten phases occurred.

II. PREPARATION OF COMPOSITES

All composites were manufactured in a one-step operation by a hot-pressing process employing graphite molds heated by induction.

The starting materials were powdery blends consisting of graphite, pitch, and the additives. Upon heating such blends under pressure, a significant primary compaction is achieved when the pitch melts between 100 and 200°C. This allows the solid particles in the system to arrange themselves in a dense pack order. After the conversion of the pitch to coke, which constitutes the formation of a carbon-bonded matrix, no further densification takes place until this pitch coke assumes a certain plasticity at temperatures above 1700°C. In this range, a secondary compaction is effected.

In general, the process was carried up to and beyond the point where a liquid phase appears in the system due to the melting of additives. As the carbon matrix is plastically compressed, the liquid phase is squeezed into the pores of the solid phase, so that, on cooling, an impervious body is formed. Also, in accordance with the equilibrium relations involved, varying amounts of the prevalent solid phase dissolve in the liquid phase. Therefore, as solubility limits are a function of temperature, processing conditions allow the adjustment of the degree of homogeneity within the composite.

The inclusion of pitch in the starting material is not necessary, but beneficial. It allows dense packing due to extensive liquefaction of the system (molten refractory additives could accomplish this only when present in sizable volume). In forming a carbon matrix, it not only provides the material with a respectable residual strength when exposed to temperatures at which a molten phase occurs, but also prevents excessive escape of molten additives by trapping them in pores. The latter benefit applies during processing, as well as during use at high temperatures.

Presses capable of producing the temperatures and pressures

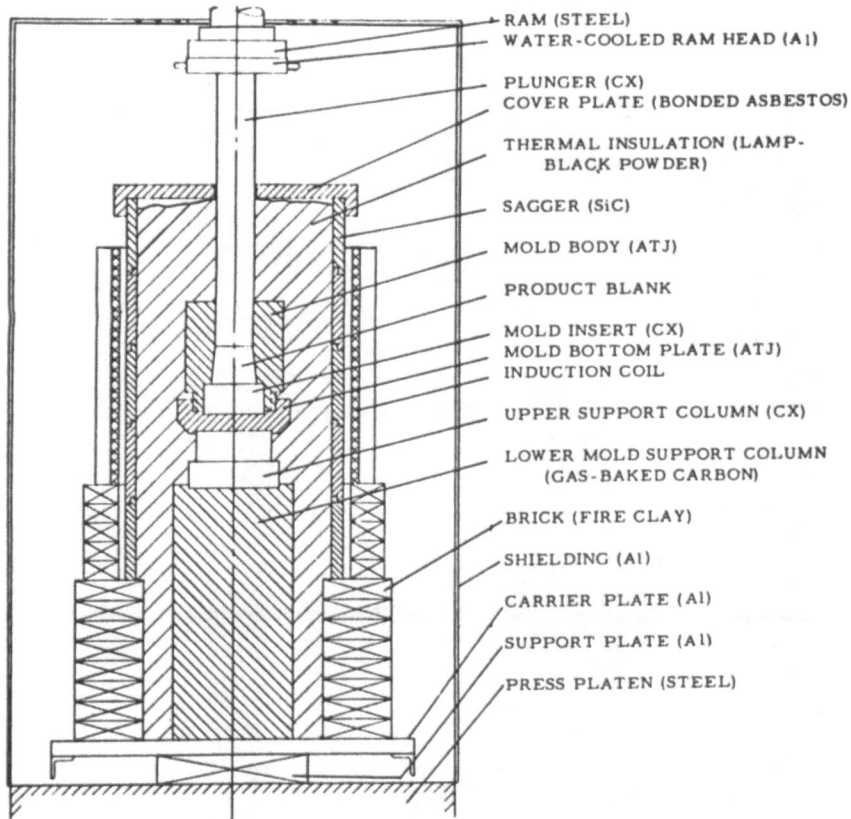

RAM (STEEL)
WATER-COOLED RAM HEAD (Al)

PLUNGER (CX)
COVER PLATE (BONDED ASBESTOS)

THERMAL INSULATION (LAMP-
BLACK POWDER)

SAGGER (SiC)

MOLD BODY (ATJ)

PRODUCT BLANK

MOLD INSERT (CX)
MOLD BOTTOM PLATE (ATJ)
INDUCTION COIL

UPPER SUPPORT COLUMN (CX)

LOWER MOLD SUPPORT COLUMN
(GAS-BAKED CARBON)

BRICK (FIRE CLAY)

SHIELDING (Al)

CARRIER PLATE (Al)

SUPPORT PLATE (Al)

PRESS PLATEN (STEEL)

Fig. 1. Hot press for composite manufacture (cross section).

required for the operations are not commercially available. It was,
therefore, necessary to construct special equipment.

The internal design of a typical press developed for this work is
illustrated in Fig. 1. The mold and its bottom plate are made ATJ
graphite. A strong consolidated lampblack material with a low
creep rate at high temperatures is used for plunger, mold insert,
and the upper support column. The mold is enclosed in fine lampblack powder which in turn is confined by a tube of clay-bonded silicon carbide. An induction coil energized by a motor–generator set
permits heating of the mold to temperatures in excess of 3000°C.

The mold support column is a sufficiently good thermal insulator
to prevent high temperatures from reaching the bottom platen of
the press. The top platen is equipped with a water-cooled head,

Fig. 2. Hot press for composite manufacture (overall view).

since the upper part of the plunger is exposed to air and has to be kept below 400°C to preclude oxidation. Aluminum shielding protects the press frame from being heated by the electromagnetic field of the induction coil. The overall setup is illustrated in Fig. 2.

III. TESTING OF COMPOSITES

Since graphite-base composites are designed to be resistant to oxidation at high temperatures, particular emphasis has been put on evaluation of this characteristic of these materials.

The test specimen is suspended in a brick-lined furnace by means of a platinum wire. The furnace is heated by resistance elements. Oxidation measurements are carried out continuously by recording the weight change of the suspended sample with a special analytical balance equipped with electrical weight compensation.

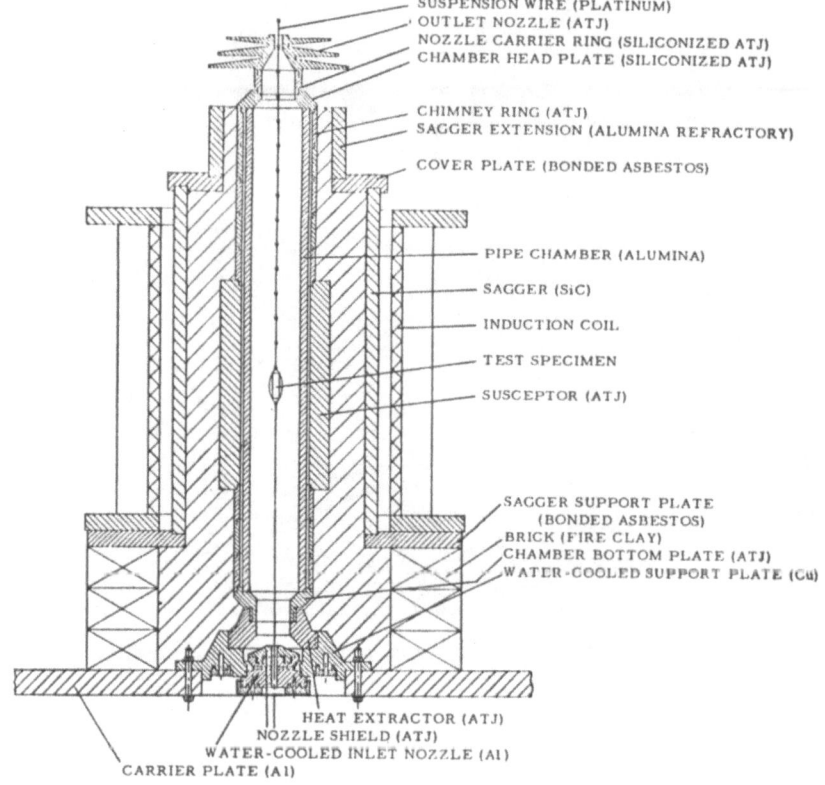

Fig. 3. Oxidation apparatus—suspension setup (cross section).

Fig. 4. Oxidation apparatus—suspension setup (overall view).

The current necessary to compensate each weight change is amplified and recorded. The outstanding feature of this apparatus is that it is capable of providing a continuous, accurate record of the sample behavior as a function of time and temperature. Unfortunately, the brick lining and the heating elements of commercial furnaces limit their temperature range to about 1400°C. To extend this range, an oxidation furnace was constructed, as shown in Fig. 3. The furnace chamber and sample are heated by radiation from a

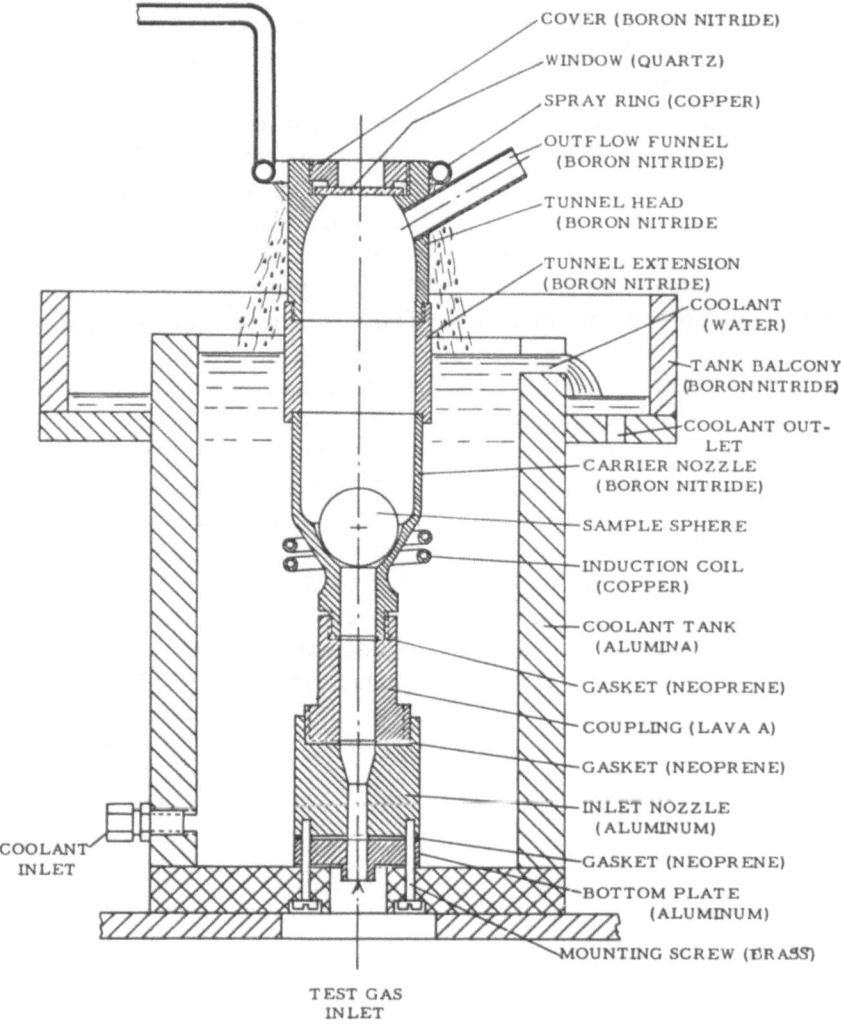

COVER (BORON NITRIDE)

WINDOW (QUARTZ)

SPRAY RING (COPPER)

OUTFLOW FUNNEL (BORON NITRIDE)

TUNNEL HEAD (BORON NITRIDE

TUNNEL EXTENSION (BORON NITRIDE)

COOLANT (WATER)

TANK BALCONY (BORON NITRIDE)

COOLANT OUTLET

CARRIER NOZZLE (BORON NITRIDE)

SAMPLE SPHERE

INDUCTION COIL (COPPER)

COOLANT TANK (ALUMINA)

GASKET (NEOPRENE)

COUPLING (LAVA A)

GASKET (NEOPRENE)

INLET NOZZLE (ALUMINUM)

GASKET (NEOPRENE)

BOTTOM PLATE (ALUMINUM)

MOUNTING SCREW (BRASS)

COOLANT INLET

TEST GAS INLET

Fig. 5. Oxidation apparatus—levitation setup (cross section).

graphite susceptor energized by electromagnetic induction. The furnace chamber consists of an impervious alumina pipe that prevents oxygen from reaching the graphite susceptor, the middle part of a column of graphite rings, which keep carbon from touching and chemically destroying the hot alumina. The chamber head plate and the nozzle carrier ring are made of siliconized graphite, since they are hot and exposed to oxygen.

The temperature of the sample is measured through a hole in the bottom plate of the furnace chamber with an optical pyrometer. Through the same hole in the bottom plate, a controlled amount of dried and purified air is fed into the chamber. An overall view of the setup is presented in Fig. 4.

The platinum wire limits the temperature range of the oxidation furnace to approximately 1650°C. Attempts to reach higher temperatures by replacing the platinum wire with rhodium wire or beryllia rods were unsuccessful, as these materials alloyed with the test specimen and disintegrated rapidly. It was, therefore, decided to construct a second oxidation apparatus in which suspension was replaced by levitation. The resulting device is illustrated in Fig. 5. A spherical sample, $1\frac{1}{2}$ in. in diameter, was levitated by the combined action of aerodynamic and electromagnetic forces, the latter stemming from an induction coil which carries out the heating. The test gas stream was confined by a silicone-coated boron nitride tunnel, the greater part of which was submerged in water to provide effective cooling. The temperature was measured and controlled by a two-color pyrometer, the sample being viewed through a quartz window in the top of the tunnel. While the test gas was generally air, heating and cooling are accomplished in argon. Uniformity of oxidation is ensured by rotating the sample electromagnetically with a tilted induction coil, as illustrated in Fig. 6.

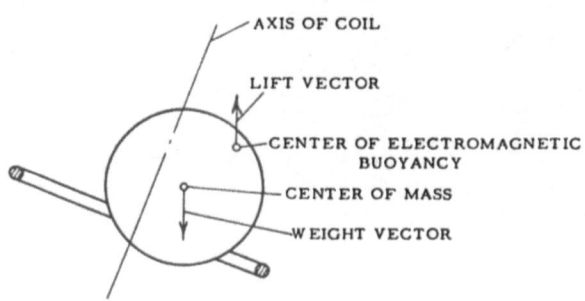

Fig. 6. Electromagnetic rotation of sample sphere.

The levitation apparatus can be used over a wide temperature ranges—from the low threshold of optical pyrometry to levels well in excess of 2000°C.

IV. RESULTS AND DISCUSSION

As the oxides of graphite are gaseous, the main purpose of additives was to provide solid oxidation products capable of encapsulating the parent material with a self-grown oxide coating. The first investigations were aimed at establishing the effects of adding to graphite well-known refractory oxide formers such as zirconium, hafnium, and columbium. It was found, however, that while these additives diminished the oxidation rate of graphite, they were not capable of halting the oxidation process since the oxides formed were neither adherent nor impervious. For this reason, attention was directed toward additives whose oxides are glass formers. It was in this second phase of composite studies where first indications of successful encapsulation were detected in the carbon–boron–silicon system. Upon exposure to air at elevated temperatures, materials in this system were found to form tight, glassy, oxide coatings which provided essentially permanent oxidation resistance up to approximately 1200°C. At higher temperatures, these coatings proved to be so liquid that they literally dripped off the surface, precluding continuous protection. To counter this liquidity, the next step was the incorporation of titanium, thus shifting the study into the carbon–boron–silicon–titanium system. Results here revealed highly satisfactory oxidation resistance up to about 1500°C, but above this temperature droplet formation and dripping recurred. It was, therefore, decided to replace titanium with zirconium, since zirconium oxide has one of the highest melting points of all oxides. The outcome of this third phase of the studies was most gratifying and culminated in the appearance on the market of the first commercial graphite-base refractory composite, grade JTA, manufactured by Union Carbide Carbon Products Division.

The oxidation behavior of JTA is characterized in Figs. 7 and 8. Figure 7 covers the range between 800 and 1400°C as investigated with the suspension apparatus, and Fig. 8 the range between 1400 and 1800°C as evaluated by the levitation technique. In both graphs, oxidation data on ATJ graphite are included for comparison. JTA exhibits excellent oxidation resistance at low, as well as high, temperatures—a result of an overlapping protective action of the

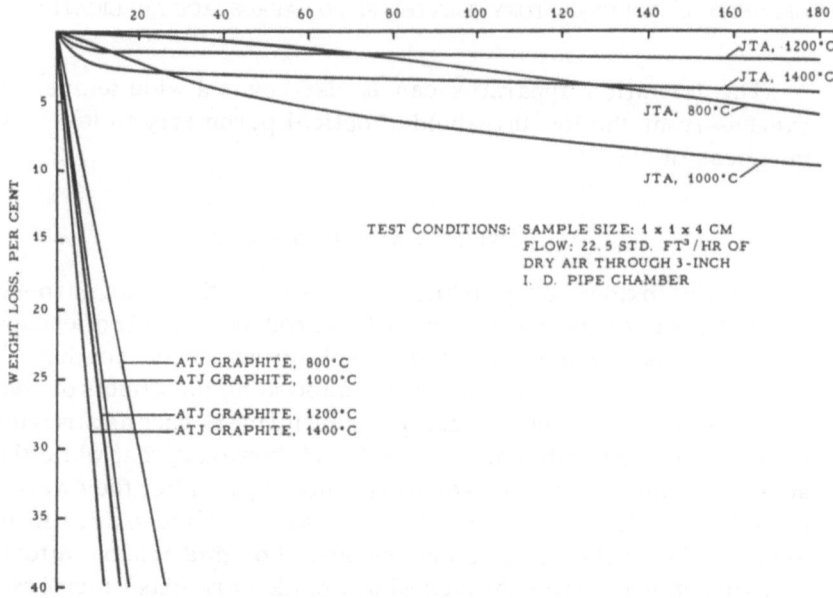

Fig. 7. Oxidation characteristics of JTA between 800–1400°C as determined in the suspension apparatus.

TIME OF EXPOSURE, MINUTES

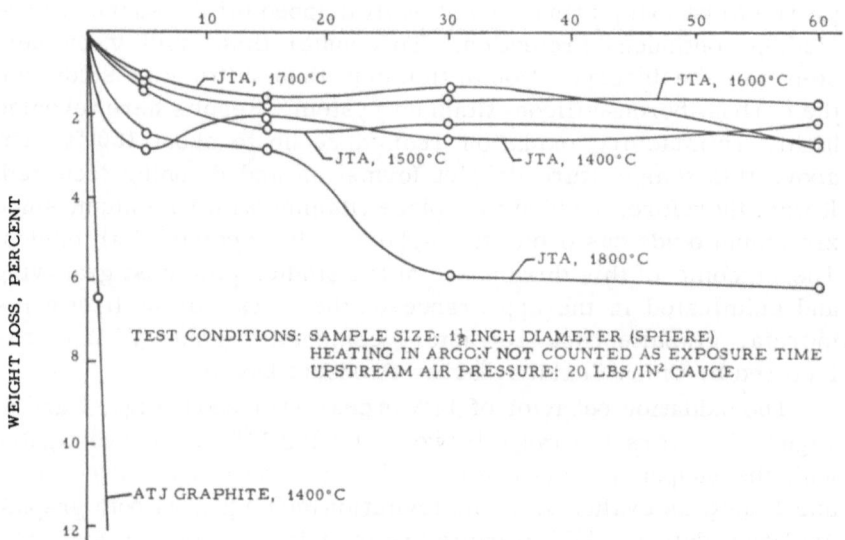

Fig. 8. Oxidation characteristics of JTA between 1400–1800°C as determined in the levitation apparatus.

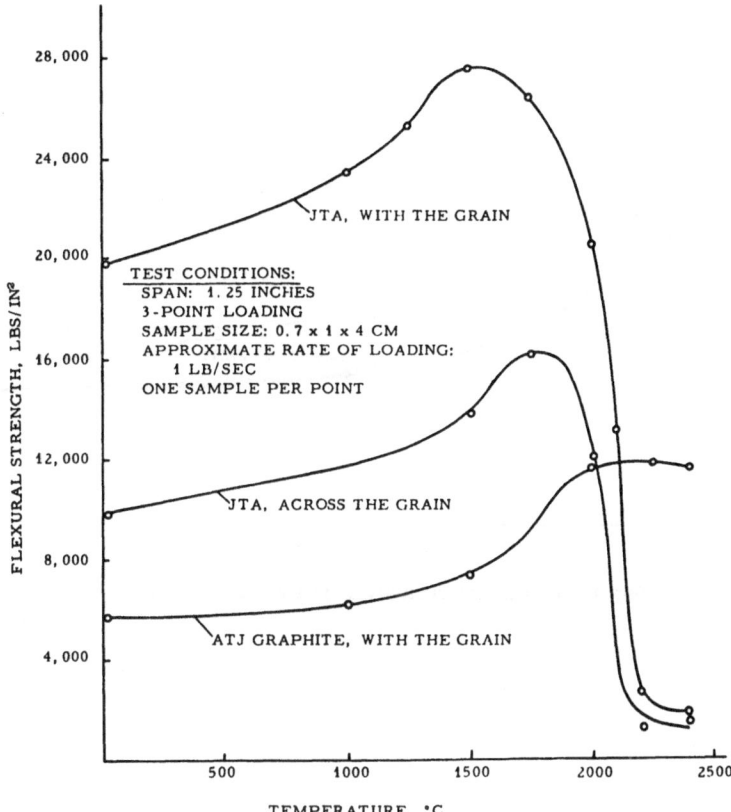

Fig. 9. Flexural strength of JTA vs. temperature.

three additives, boron, silicon and zirconium, with B_2O_3 being predominantly effective at low temperatures, SiO_2 at intermediate temperatures, and ZrO_2 at high temperatures.

For structural applications, the usefulness of JTA is enhanced by outstanding strength at temperatures up to 2000°C, as illustrated in Fig. 9. Thermal shock resistance is provided by good thermal conductivity and moderate thermal expansion, as depicted in Figs. 10 and 11, respectively. The combination of these desirable properties with oxidation resistance has permitted application of JTA under severe conditions, such as those found in rocket motors and other structural re-entry regimes. At temperatures above 1800°C, the interaction between the additives in JTA no longer prevails, in that boron becomes a detrimental agent. B_2O_3 develops sufficient

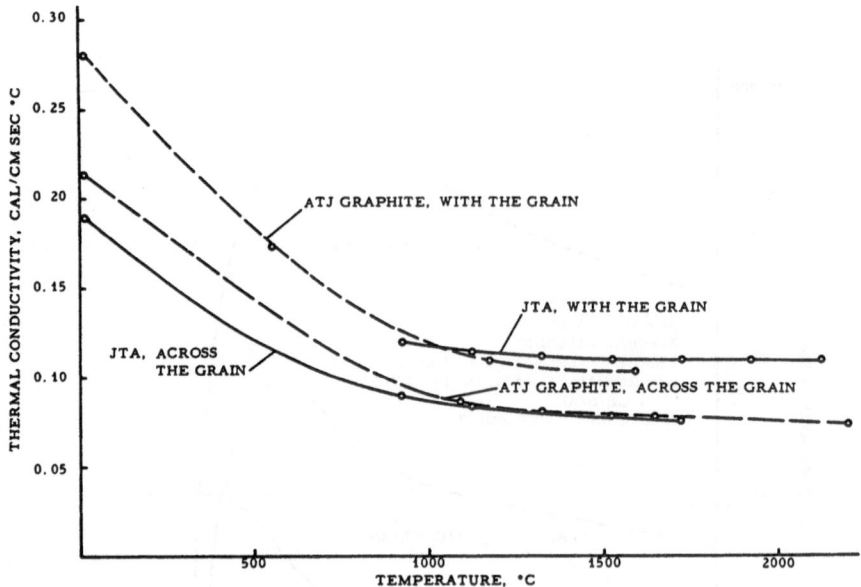

Fig. 10. Thermal conductivity of JTA vs. temperature.

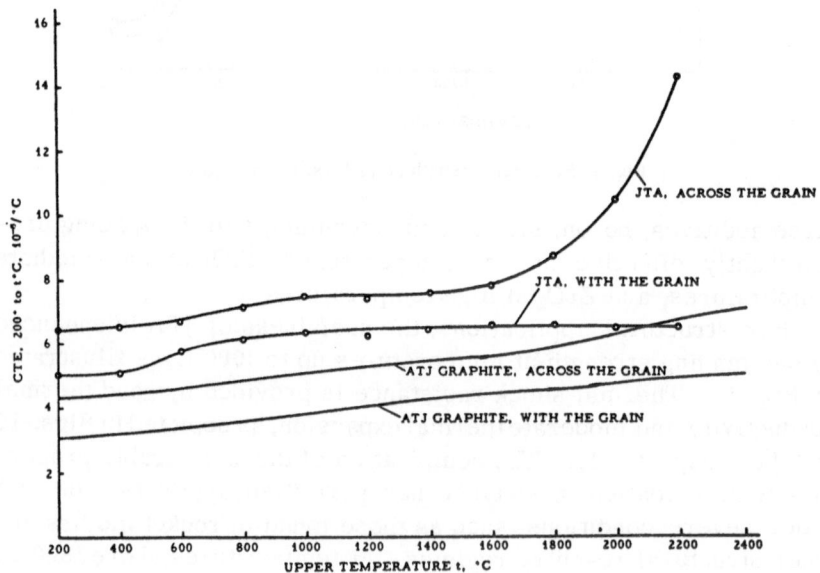

Fig. 11. Coefficient of the thermal expansion of JTA vs. temperature.

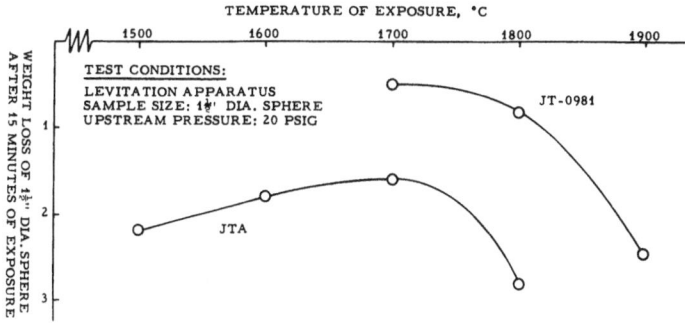

Fig. 12. Comparison of JTA with a novel high-temperature composite; weight loss of a 1½-in. diameter sphere after 15 min of oxidation vs. temperature of exposure.

activity to readily evaporate from the coating and also supports carbon/oxide reactions in the interface between the coating and the parent material. For this reason, composites designed for very high temperatures do not contain boron. While this deprives them of good oxidation resistance at low temperatures, the benefit of the absence of boron at high temperatures is clearly established, as illustrated in Fig. 12, where JTA is compared to one of the newer high-temperature composites of equal carbon content.

Developmental work is continuing, the ultimate goal being a graphite-base refractory composite with permanent oxidation resistance at temperatures in excess of 2000°C. With the soundness of the composite principle established and with extensive carbon/oxide reaction studies at high temperatures as fundamental support, it is anticipated that it will soon be possible to tailor graphite-base composites to meet the specific requirements of a wide variety of applications.

Glass Microtape

Richard A. Humphrey

DeBell & Richardson, Inc., Hazardville, Connecticut

Progress toward the making of research specimens of an entirely different filament-wound structure is described. Glass-fiber-reinforced plastics have been shown by others to have mechanical properties superior to most metals at cryogenic temperatures. However, the resin phase is permeable to hydrogen.

The objective of this work is to wind low-permeability shells as suitable containers for cryogenic liquids. The plan is to draw continuous glass microtape about 0.0005 in. thick by 30 to 50 times as wide, and subsequently filament-wind the microtape into thin-walled cylindrical composites.

The preform attenuation method which entails drawing a filament from a preform as it passes slowly downward through a small furnace is used to draw microtape from a strip of sheet glass. No dies are used. Only through careful study of the dynamics of the drawing process has a furnace design been developed which produces a flat microtape without an edge bead. A typical continuous microtape is about 10 μ thick and 400 μ wide, or 0.0004 by 0.016 in. The winding of microtape into solid, perfectly packed structures has developed into a separate research task. A commercial textile spooler has been modified to provide the fine, uniform traverse necessary for perfect side-by-side placement of microtape into a helically wound, virtually all-glass, resin-bonded cylinder.

I. INTRODUCTION

Glass-fiber-reinforced plastics have good mechanical properties at cryogenic temperatures. This feature made these composite materials attractive for fabricating cylindrical, cryogenic, liquid-propellant containers. Unfortunately, the minute hydrogen molecule can readily permeate through the resin phase, which indicated a liner would be necessary.

The problem, therefore, is to develop a wide, thin, flat microtape which should minimize permeability through a filament-wound structure. This would circumvent the need for a liner and might, at the same time, provide extra properties not available with round filaments or strands presently used in filament winding.

Glass microtape is a continuous, solid tape of microscopic dimensions. Figure 1 attempts to illustrate this definition. Most commercial glass fiber is about 0.00037 in. in diameter. A microtape has been made consistently on a laboratory scale at

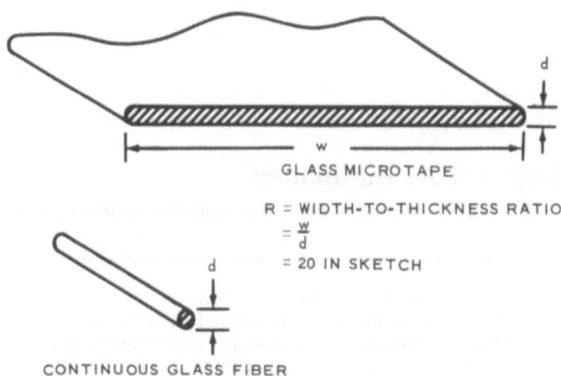

Fig. 1. Comparison of glass microtape with glass
fiber.

DeBell & Richardson, Inc. that is 0.0004 in. thick and 0.015 in. wide,
a width-to-thickness ratio of about 35.

II. THEORETICAL ADVANTAGES

An examination of leakage paths through composite structures is
important. Permeability through a glass-fiber-reinforced structure
is inversely proportional to the leakage path length, and approximately proportional to its cross-sectional area and the number of
leakage paths per unit area. Permeability of glass to gases is
negligible; only the resin phase needs to be examined.

Round fibers, even in perfect hexagonal packing as shown in
Fig. 2, present a leakage path length L of $\pi/3$ times (only 1.05
times) the wall thickness t of the vessel under consideration.

It is easy to see how microtape can extend the leakage path
length. Consider a microtape whose width-to-thickness ratio R is

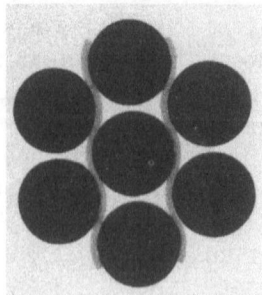

Fig. 2. Diagram of leakage path through round
filaments — hexagonal packing.

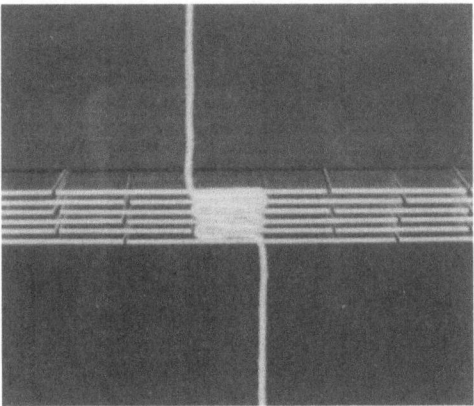

Fig. 3. Model of microtape composite cross section
with perfect overlap showing leakage path.

50 and is wound in flat layers in close juxtaposition to one another.
If this is staggered in winding so that a microtape is centered
exactly over the joint in the underlying layer, as shown in the model
(Fig. 3), the leakage path becomes

$$L = \left(\frac{R}{2} + 1\right) t$$

$$L = \left(\frac{50}{2} + 1\right) t = 26t$$

This assumes the number of layers of microtape is large, as is
the case with the thin microtape under consideration. Thus, in the
same wall thickness, the 50 : 1 microtape presents a leakage path
which is 25 times as long as that for perfectly packed round fibers.
Even a $\frac{1}{3}$ overlap (Fig. 4) provides a path over 19 times that for
round fiber. A purely random winding of microtape side by side
will provide a leakage path which is equivalent to a $\frac{1}{4}$ overlap. The
path length then becomes

$$L = \left(\frac{R}{4} + 1\right) t$$

$$L = \left(\frac{50}{4} + 1\right) t = 13.5t$$

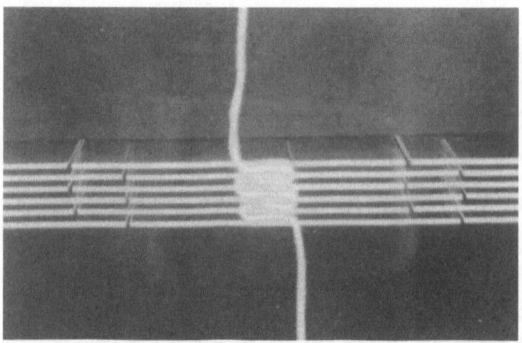

Fig. 4. Similar model, but with one-third overlap showing
shorter leakage path.

This is still over 12 times the path length of a perfectly packed round filament composite.

Another difference found between a round fiber composite and one made from microtape is the cross-sectional area of the leakage path. Going back to the well-packed round filament composite (Fig. 2), the cross section of the leakage path is related to the size of the filaments. For discussion purposes, consider a standard commercial fiber of 0.00037 in. The area the gas molecule "sees" is at right angles to the one seen when studying a section through the ends of the filaments. One of its dimensions is the circumference of the cylindrical vessel and is the same for microtape. The other dimension will average no less than $\frac{1}{8}$ the filament diameter or 0.000046 in. To arrive at this estimate, the tricorns where three round filaments are nearly in contact were calculated as equilateral triangles, and the average altitude of such a triangle was found to be $\frac{1}{8}$ the diameter of the filaments.

With a good quality filament winding of microtape, on the other hand, a resin layer between flats of microtapes of less than $\frac{1}{4}\,\mu$, or less than 0.00001 in., has been measured.

A more realistic appraisal of existing conditions in winding of

Fig. 5. Diagram of leakage path through round
filaments — square packing.

round filaments would be to assume an approach to square packing (Fig. 5) which gives a path length equal to the wall thickness and an average resin layer through the quadricorns approaching $\frac{1}{3}$ the diameter of the filaments, or 0.00012 in.

Thus, a comparison, on paper, of permeability of 50 : 1 microtape composite with optimum filament winding practice shows the path length to be at least 12 times longer for microtape and a resin layer over 12 times thicker in round filament windings. The number of leakage paths through a unit area of wall is 50 times greater in a round filament-wound structure than it is with 50 : 1 microtape. Therefore, the permeability of a microtape structure is predicted to be less by the following factors because of the different path characteristics:

Length	Thickness	Frequency	
$\dfrac{1}{12}$ $\cdot$	$\dfrac{1}{12}$ $\cdot$	$\dfrac{1}{50}$ $=$	$\dfrac{1}{52,200}$

In other words, the permeability of a microtape structure might be only $1.9 \cdot 10^{-5}$ times that for an ordinary filament-wound vessel.

III. DEVELOPMENT OF MICROTAPE COMPOSITES

The microtape development program has two distinct parts: One involves learning how to produce a flat, thin tape; the other concerns the perfect placement of the tape in a composite. One must have a suitable microscopic tape before it is possible to apply it.

To obtain the requisite flat microtape to achieve the desired ultrathin resin layers, it has been clear from the outset that little

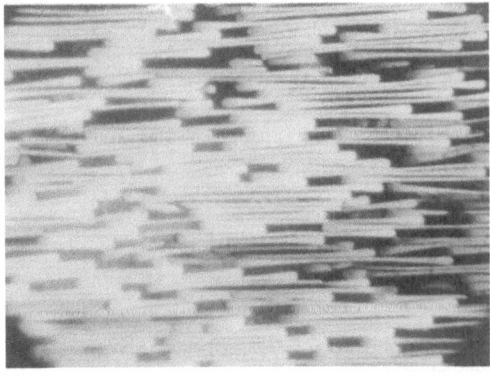

Fig. 6. Photomicrograph (60 ×) of polished section through microtape with edge beads.

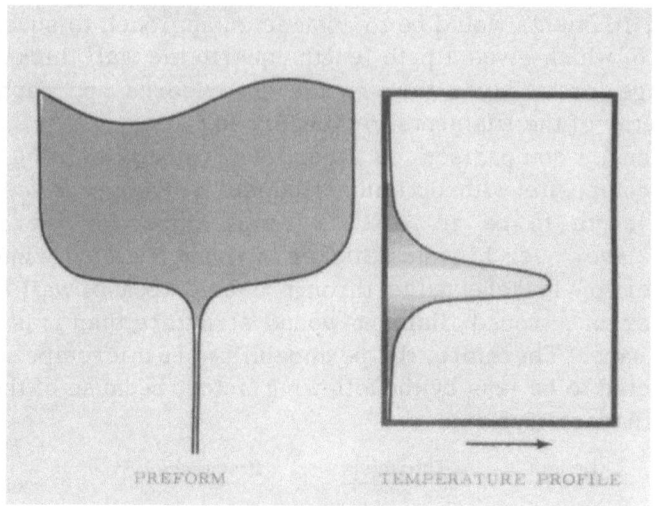

Fig. 7. Diagram of preform in thin horizontal hot zone.

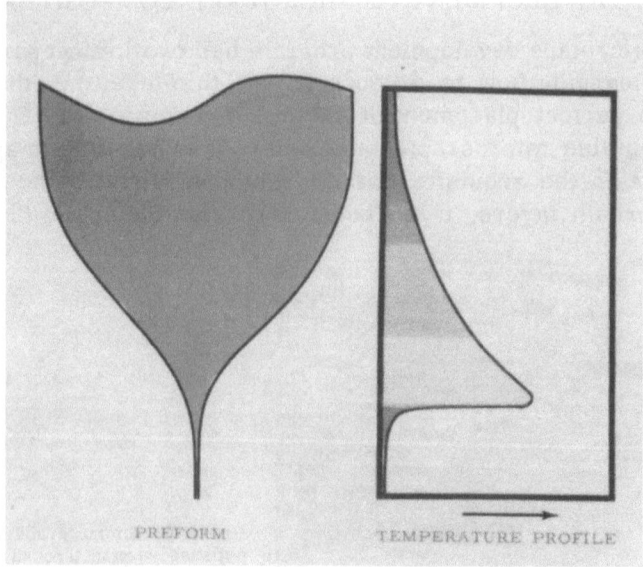

Fig. 8. Diagram of preform in improved furnace (long preheat and sharp cutoff).

or no bead at the edges of the tape can be tolerated. There is today a so-called ribbon glass, somewhat larger than microtape, made in this country from a platinum bushing using a process related to that used for making the vast majority of round-filament glass fibers. This process relies upon the beads to form the flat center section [1].

The success in our laboratory with the older preform attenuation process led to its use on this problem. However, first attempts to make high ratio microtape from a wide preform also yielded edge beads (Fig. 6). A study of the dynamics of the process showed that the short vertical hot zone which had proved satisfactory for three-dimensional fiber attenuation was causing the critical edge portion of the preform to remain in the maximum temperature zone too long during attenuation (Fig. 7). The abrupt narrowing of the preform shows this. The graph beside it illustrates the vertical temperature profile.

Several modifications to the furnace were made to overcome this beading. The heating elements were spaced farther from the flat sides of the preform; the slots through which the heat could "shine" onto the preform were enlarged upward to let the heaters "see" the preform longer, providing a more gradual preheat; and, lastly, a cooler was inserted from the bottom of the furnace up to the hottest point the fiber reached, about on the level of the heating elements, causing an air quench at the moment attenuation was completed. Figure 8 shows the much more gradual attention of the sheet

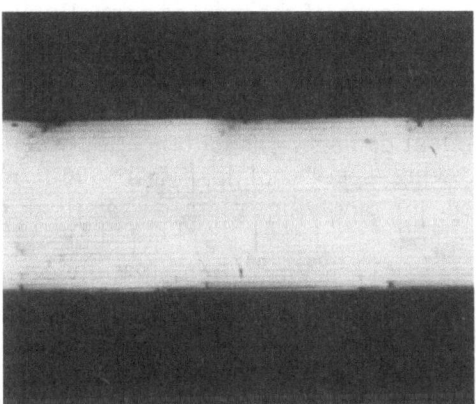

Fig. 9. Photomicrograph (60 x) of flat microtape
cross section through wound hoop.

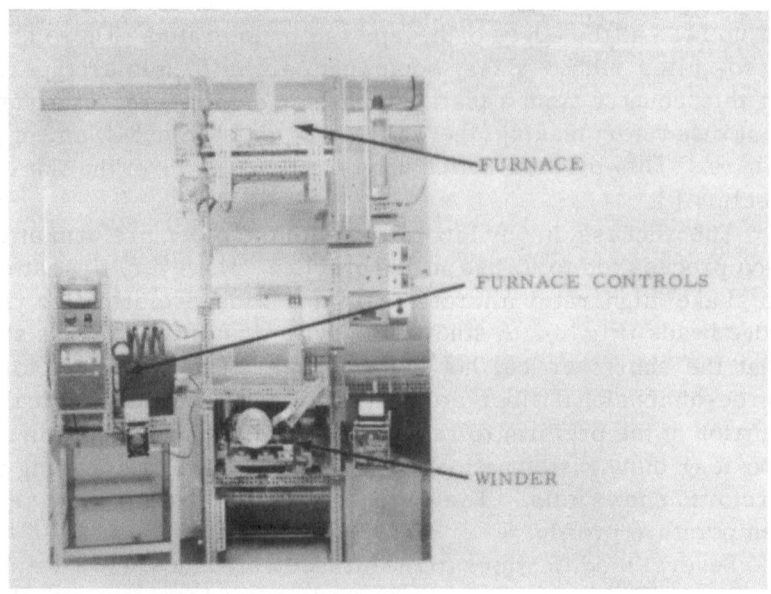

Fig. 10. General view of furnace and winder.

glass preform and compares it with the extended asymmetrical vertical temperature profile. Figure 9 is a photomicrograph of a cross section through a hoop wound from the improved flat microtape. This tape has a width-to-thickness ratio of about 37.

A view of the furnace and winding equipment for making microtape is shown in Fig. 10. The mechanical feeder is on the floor above and is capable of lowering the preform at a controlled rate. Figure 11 shows a closeup of the furnace partially dismantled. The position of the heating elements and the cooler is indicated.

A wet winding process directly under the furnace promises to make the best microtape composite. Immediate winding provides minimum mechanical damage. The attenuation process is presently done at low drawing speeds of less than 500 ft/min, which is a reasonable speed for applying resin as well as tensioning and guiding the microtape.

Winding the tape in a perfect, fine pitch thread onto a cylinder is the other part of the development. The 0.015-in. wide microtape requires a placement accuracy of less than ±0.0005 in. Preliminary hoops were made with a winder which incorporated one variable-speed motor for the winder and another for driving an endless thread traverse mechanism. Coordination of these two devices required innumerable trials and microscopic examination of the lay. Once coordination was achieved, it was found that minor variations in the traverse mechanism would cause gaps in the winding or, more catastrophic, piling up of one filament on the next.

The final design of the cylinders will be based on the hoop tensile strength as well as the tensile strength parallel to the axis of the cylinder of the microtape composite. With the large areas of pure shear imposed by the wide flat microtape, it is expected that the tensile strength in the plane of the tape and normal to its length will be unusually high for a filament-wound structure.

Up to the present time, only window glass and sheet Pyrex glass have been attenuated into microtape, primarily because of their ready availability in sheet form. Other factors influence the choice of the base glass. Glasses with short working ranges, that is, steep slope of viscosity versus temperature, require overly precise control of furnace temperature and gradient. The traditional glass used for glass fiber, E glass, originally developed for electrical application since it contains only small amounts of alkali, has a very short working range, emphasizing still another factor in choosing the glass.

When held at or near the forming temperature for any length of time, E glass, even more than common soda-lime glass, devitrifies or crystallizes partially, creating stress concentrations which yield weak fibers. It has been found by Holloway [2,3] that when a glass

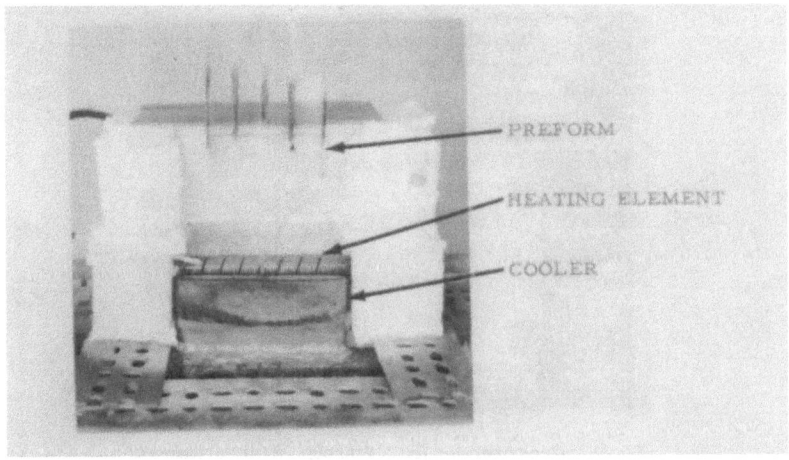

Fig. 11. Closeup of partially dismantled furnace.

fiber is drawn from a glass rod, which is in essence the preform attenuation process, any particulate matter on the surface of the rod, such as dust or ash from fingerprints, can pass through the furnace and remain attached to the surface and likewise cause a stress concentration leading to a weakened fiber. However, he has also demonstrated that glass rods which have previously been scrupulously cleaned and acid-polished can be attenuated into high-strength fibers.

One important consideration in arriving at a suitable composition for making glass fibers is resistance to attack by moisture, an essential property because of the large, exposed surface area. E glass is highly resistant to moisture attack. Because of the low permeability of a microtape structure, it may be found that only nominal resistance to moisture attack will be required which may make it permissible to use window glass.

IV. FUTURE APPLICATIONS

It has been predicted that structures with tapelike reinforcements and thin, flat resin layers should withstand higher compressive loading than ordinary composites wound with round fibers. The theoretically high glass contents should yield an elastic modulus approaching that of the glass without exhibiting the catastrophic failure of monolithic, brittle materials. The particular application

Fig. 12. Cross section (60 ×) through multipartitioned hollow fibers.

for which this microtape is being developed primarily requires composite tensile strength, since the cylinder is required to maintain moderate internal pressure. Interest has been shown in hollow glass reinforcements for structures to be exposed to high compressive loading, such as vessels designed for deep submergence.

Figure 12 is a photomicrograph of a polished section through some fibers developed for NASA Headquarters [4]. This fiber shape demonstrates one that would have thin resin layers in filament-wound structures and hollowness to contribute microstiffness.

Another approach to a hollow tape is the use of a row of round tubes for a preform. This leaves a corrugated surface which might improve the interlaminar shear strength and the peel strength. A number of different-design, hollow, tapelike filaments have been made, and there seem to be few limitations on what designs could be made. At this time, the potential applications need to be brought to the attention of the laboratory for solution.

V. CONCLUSIONS

Glass microtape is a newly developed form of the well-known material—fibrous glass. Since its development is still in progress and since the techniques for winding it are still being refined, it has not yet been tested in the application for which it is presently being investigated. The problems still to be solved are difficult, but a way to overcome them has been chosen which should provide good results. There is good reason to expect microtape to provide a satisfactory solution to the requirements for containers for cryogenic liquids. Quite possibly filament-wound glass microtape could accomplish other related aerospace jobs in that it is a material whose properties approach those of solid glass without having the disadvantages of a monolithic glass object. In addition to these low-temperature applications, there should be many others which could be filled better by microtape than by present materials, not only for aerospace but also for a wide variety of structural uses.

ACKNOWLEDGMENT

This work is being supported by NASA Lewis Research Center under Contract NAS 3-3647. Thanks are due to Messrs. Jack Esgar, Robert Hickel, and Morgan Hanson, all of Lewis Research Center, for their suggestions and encouragement. The microtape program is supervised by Mr. William J. Eakins. The skill and ingenuity of technician Lewis Heath has been an important factor in the successful development of microtape.

REFERENCES

1. D. Brown, Corning Glass Works, U.S. Patent 2,422,466.
2. D. G. Holloway, "Strength of Glass Fibers," Phil. Mag. 4 (series 9): 1101-6 (1959).
3. D. G. Holloway and P. Hastilow, "High Strength Glass," Nature 189: 385-386 (1961).
4. W. J. Eakins and R. A. Humphrey, "Studies of Hollow Multipartitioned Ceramic Structures," NASA Contractors Report NASA CR-142, December 1964.

Material Design Concepts for Uncooled Nuclear Rocket Nozzles

S. R. Locke and R. L. Ahearn

Martin Company, Denver, Colorado, and Orlando, Florida

An uncooled nuclear rocket nozzle would provide significant weight, cost, and reliability advantages. Current data on materials indicate that composite materials systems would be capable of successful operation in the extreme environments imposed on nuclear rocket nozzles. This paper presents a four-basic-composite-materials-systems concept which has been developed under a preliminary program sponsored by Martin-Orlando.

I. INTRODUCTION

Considerable design analysis has been made in the past on regeneratively cooled nuclear rocket nozzles. However, only recently has interest centered on uncooled, or more accurately, radiation-cooled nozzles. This is because refractory material combinations appear to be capable of meeting the unique requirements of the nuclear rocket nozzle environment. The greatest potential advantage of the uncooled nozzle is greater system reliability. This paper will not compare the two nozzle systems as to mission objectives, but will present, in detail, several uncooled nozzle materials systems that would be capable of successful operation on long flight-time, interplanetary missions.

A study of the feasibility of uncooled nuclear rocket nozzles has not appeared in the literature. The approach to this presentation is a logical consideration of all environmental factors in material selection. This environment includes the anticipated heat fluxes, pressures, mass transfer rates, chemistry, and the unique radiation exposures found during operation. However, for the purpose of this paper, an arbitrary nozzle environment was established. The various materials systems are analyzed on the basis of thermodynamic, stress, and chemical reaction standpoints. The material systems are designed so that they overcome the inherent shortcomings associated with the refractory materials taken individually, and in fact complement each other's functions. This has been

287

TABLE I

Properties of Some Representative Refractory Materials

Material	Melt. temperature (F°)	Density (g/cc)	Thermal conductance	Erosion resistance	Chemical reaction	Thermal structure	Mech. 4500°F	Fabric-ability	Miscellaneous
Tungsten	6100	19.3	Good	Excellent	Volatile oxide	Poor	Good	Poor	Poor weld, very brittle
Rhenium	5740	20.5	Good	No data	Oxidizes	No data	No data	Good	Expensive
Tantalum	5430	16.6	Good	Poor	Reactive	Good	Fair	Excellent	Ductile
Molybdenum	4710	10.2	Good	Fair	Volatile oxide	Fair	Poor	Fair	Poor weld and ductility
Columbium	4380	8.6	Good	Fair	Reactive	Fair	Poor	Good	
W-Ta alloy	6100–5430	16.6–19.3	Good	Good	Oxidizes	Fair	Good	Fair	
W-Mo alloy	6100–4710	10.2–19.3	Good	Good	Oxidizes	Fair	Good	Fair	
W-Ph alloy	6100–5740	19.3–20.3	Good	Good	Oxidizes	Fair	Good	Poor	

Material									
Molded graphite	6600^S	2.25	Good	Fair	Oxidizes	Good	Fair	Excellent	
Pyrolytic graphite	6600^S	2.25	"a" excellent	Excellent	Oxidizes	Good	Excellent	Poor	Available in thin section
Pressure baked graphite	6600^S	2.25	Good	Excellent	Oxidizes	Good	Good	Excellent	
4 HfC TaC	7177	14.2	No data	No data	No data	Poor	No data	Poor	⎫
4 TaC ZrC	7110	13.3	No, data	No data	a	Poor	11,380^c	Poor	⎪ Limited availability
HfC	7030	12.7	13*	No data	a	Poor	12,640^c Slight creep	Poor d	⎬
TaC	7020	14.7	23*	Fair	b	Poor	17,600^c Slight creep	Poor d	⎭
ZrC	5750	6.4	27*	Poor	a	Poor	2500^c	Poor	
NbC	6330	7.8	15*	No data	b		2800^c Slight creep	Poor d	

*Btu/hr/ft/°F at 3800°F.

a Stable in H$_2$ to 3630°F.

b Attacked by H$_2$ above 2730°F.

c Psi in bending at 3600°F.

d At 4000°F.

TABLE II

Material Properties

Material	Strength tensile Ksi	Strength comp. Ksi	Expansion ($N \times 10^{-6}$)	Thermal conductivity Btu/ft^2-hr-in/°F	Modulus of elasticity psi ($\times 10^6$)	Relative thermal shock resistance	Relative abrasion resistance
Tungsten	200		2.4 11/11°F	1397	59 16 at 390°F	Good	Good
Tantalum	60		3.7 11/11°F	377	27	Good	Good
Molybdenum	100		2.7 11/11°F	967	43	Good	Good
ThO$_2$	14	220	25°-800°C 9.5×10^{-6}	Very low	17.9×10^6	Fair	Good Reduced by carbon at high temperatures
ZrO$_2$	17.9	302	0-1400°C 5.0×10^{-6}	2400°F 14.3	24.8×10^6 25	Fair	Good Forms carbides at high temperatures
Al$_2$O$_3$	35.8	412	250-800°C 8.5×10^{-6}	2400°F 30.0	52.4×10^6	Good	Good Not attacked by graphite
Graphite	2.4	10	See curves	400°F-730 2000°F-220	2.2×10^6 (50% higher at 3600°F)	Very good	Poor

accomplished by utilizing the refractory material in sizes and shapes such that they exert their own behavior and still overcome their shortcomings. We find that the nuclear rocket nozzle environment is unique in that, in some respects, it is easier to cope with than the environments of solid or liquid propellant rockets, while in other respects the environment is much more demanding on the materials. For example, the nuclear flame temperature is only about 4000°F as compared to 6500°F in the solid propellant motors. Also, the nuclear rocket operates in a vacuum, which solves many of the most severe oxidation problems of conventional flight regimes. However, flight durations of from 0.5–100 hr at 4000°F provide a severe design restriction for a nuclear rocket nozzle. In selecting a nozzle materials system to operate under these conditions, we are faced with the choice of finding a material with adequate structural strength and creep resistance at operating temperatures, or else insulating a material which has adequate structural strength at some lower temperature. Because of the long operating-time requirement for the nuclear rocket nozzle, a thermal insulation must be selected that is stable at the operating temperature and that is compatible with the metal substructure and the working fluid, which is hydrogen.

II. MATERIAL CONSIDERATIONS

We find that there are many materials, such as tungsten, molybdenum, tantalum, graphite, and certain mono and binary carbides, which maintain their structural integrity under an intense radiation field and at high temperatures. But when considering a material system for an uncooled nuclear rocket nozzle that is to operate at 4000°F or higher for a considerable length of time, we are faced with the problem of finding materials with adequate structural strength, creep resistance, and neutron resistance at the given operating temperatures. The homogeneous materials with melting temperatures above 4000°F or even 3000°F all have some critical shortcomings that preclude their use by themselves for this application. It can be seen that every material listed in Tables I and II has one or more shortcomings that prevent it from being used when exposed to a nuclear rocket's extreme operating condition.

The major limitation of these materials for service above 4000°F is melting temperature. Unfortunately, the vast majority of materials suitable for application as nuclear rocket nozzles have melting temperatures below 4000°F. The materials with suitable melting points above 4000°F are also limited. Many have low room-temperature ductility, thermal shock, and oxidation resistance, and poor high-temperature strength.

However, close scrutiny of Table I reveals that a combination of acceptable properties could be realized by combining those materials which have desirable properties into an overall composite system which may overcome their undesirable characteristics.

Recent developments in the field of refractory composite materials for solid rocket missiles have provided these combinations and have produced successful material systems where each material component complements the other's function [1]. This has been accomplished by utilizing materials in sizes and shapes such that they could exert their own behavior and still overcome their shortcomings. In order to accomplish this, careful attention was paid to the physical and mechanical properties of each member. In particular, such properties as thermal coefficient of expansion, tensile and compression strength at various temperatures, modulus of elasticity, surface chemical and physical compatibility, and oxidation or erosion resistance, or both, were taken into strict account.

Because of the long operating time the nuclear rocket nozzle will experience, a thermal insulation must be selected that will be stable at the operating temperature in addition to being compatible with the metal substructure. An organic base insulation, such as asbestos phenolic, which is commonly used with solid-propellant rocket nozzles would be inadequate for the nuclear nozzle operating condition. Therefore, for this application, we must look to the refractory or ceramic insulators which would be compatible with any metal reinforcement or substructure or both.

As long as the ceramic, such as a refractory oxide, does not break down at its operating temperature so that oxygen is not available for embrittlement of the reinforcing and substrate structures, we can avoid an antioxidation coating of the metal. Since the nuclear rocket nozzle operates in a vacuum environment, oxidation protection of the external structure is not necessary.

Generally, graphite is not a good insulator. However, we find that some special forms of graphite (pyrolytic) are available that make it possible to use this material as an insulator, so long as the surface temperature of the graphite in contact with the metal substructure does not exceed the temperature at which the metal

TABLE III

Material Properties

Property	Tungsten	Tantalum	Molybdenum
Highest working temperature in which after 100 hr the evaporation loss does not exceed 1% in vacuum of 10^{-4} mm Hg	4640°F	4350°F	3470°F
Rate of evaporation in vacuum of 10^{-4} mm Hg mg/cm/hr, at			
3146°F	1.3×10^{-4}	5.9×10^{-6}	3.1×10^{-4}
3506°F	5.3×10^{-8}	5.9×10^{-8}	3.6×10^{-2}
3866°F	4.6×10^{-4}	1.1×10^{-4}	180 -
4226°F	1.4×10^{-2}	2×10^{-1}	Too high for
4586°F	2.7×10^{-1}	2.5	long time use
Stability towards insulation and atmospheres:			
Graphite	Carbide formation beyond 2550°F	Carbide formation beyond 1830°F	Carbide formation beyond 2190°F
Al_2O_3	Stable ———————————up to 3470°F ———————————		
ZrO_2	Stable to 2910°F	Stable to 2910°F	Stable to 3470°F
ThO_2	Up to 4000°F	Up to 3470°F	Up to 3400°F
Hydrogen (dry)	Stable up to melting point	Hydride formation at N1000°F beyond that, stable up to melting point	Stable up to melting point
Refractory carbides and nitrides	Unstable at 4000°F		
Vacuum < 10^{-2} mm Hg	Stable up to 3630°F	Becomes brittle through getter action of any trace of H_2,C,O_2, or N_2. N$_2$ is removed in high vacuum.	Stable up to 3090°F

(Continued)

TABLE III (cont.)

Material Properties

Property	Tungsten	Tantalum	Molybdenum
$< 10^{-4}$ mm Hg	Strong evaporation above 4350°F	Strong evaporation above 4000°F	Strong evaporation above 3270°F
Recrystallization			
Ductile-brittle	Brittleness occurs after long heating	No brittleness occurs after long heating	Brittleness occurs after long heating
Transition temperature	645°F	-320°F	85°F

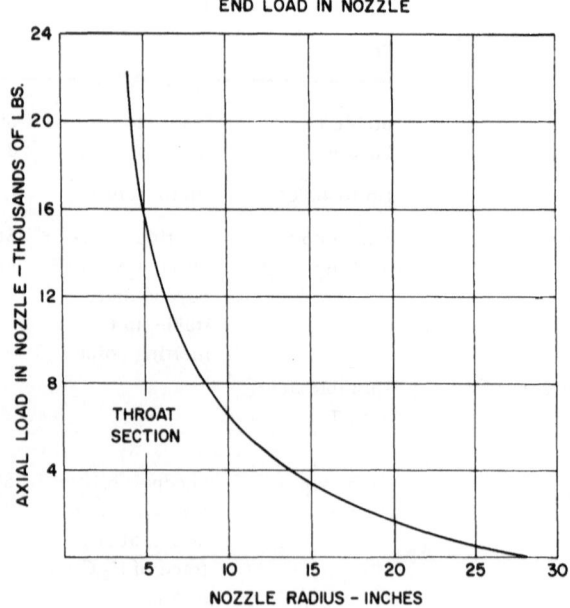

Fig. 1. End load in nozzle.

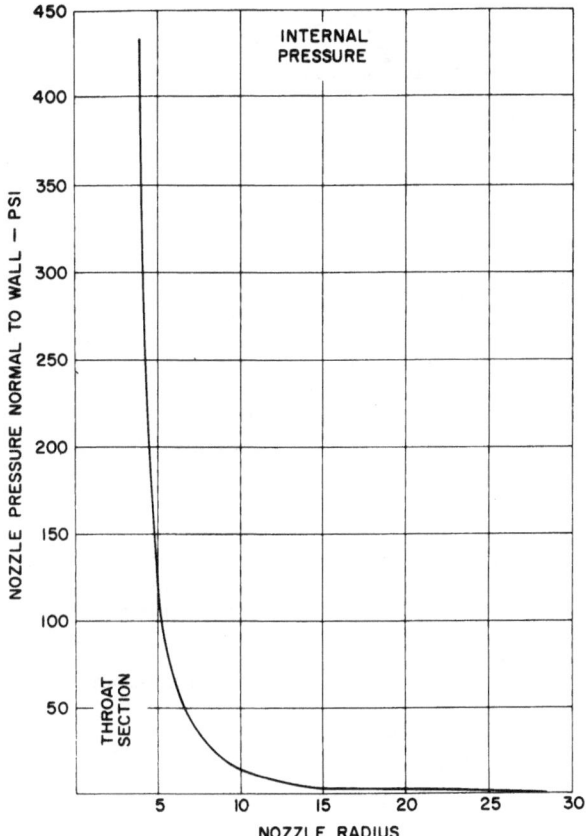

Fig. 2. Internal pressure.

could be weakened because of carbide formation. These temperatures at which carbides form are shown in Table III. Table III also shows other pertinent properties of candidate refractory materials that must be considered for a nuclear rocket nozzle, such as a reaction of these materials with hydrogen.

III. DESIGN CONCEPT

In order to design an uncooled nozzle structure, it is first necessary to define the total anticipated environment. The total environment would include the heat fluxes, pressures, mass transfer rates, and chemicals involved, coupled with any unique nuclear radiation exposures the nozzle encounters during operation. For the

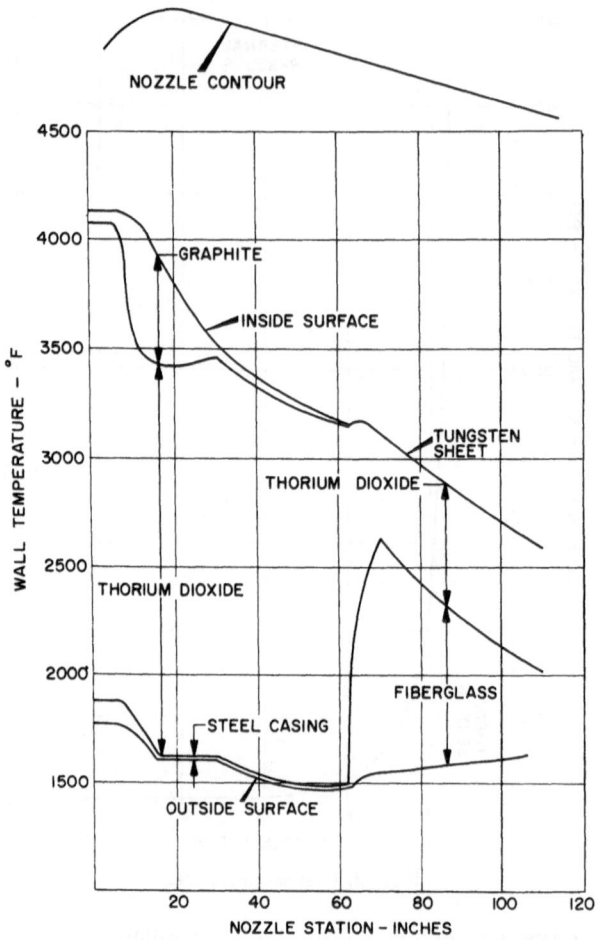

Fig. 3. Predicted temperature distribution in the nozzle wall — concept A.

purpose of this paper, an arbitrary nozzle environment has been established. The designed rocket nozzles are suitable for a nuclear rocket to be used in accomplishing interplanetary missions. It is to be uncooled and must be capable of intermittent operation for periods up to 30 min, for a cumulative time of 100 hr. Gaseous hydrogen is to be the propellant medium. In addition, the following design conditions or limitations are assumed: (1) nozzle throat diameter—8 in., (2) expansion ratio—50, (3) exit diameter—57 in., (4) exit divergence angle (total)—30°, (5) chamber pressure—800 psi, (6) gas total temperature—4000°F, (7) mode of heat dissipation—

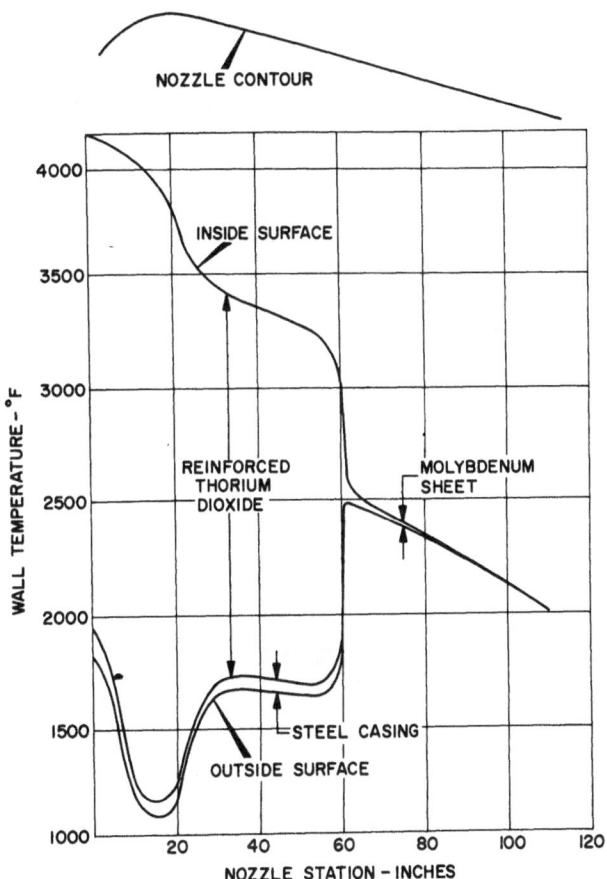

Fig. 4. Predicted temperature distribution in the nozzle walls — concept B.

radiation to space, and (8) weight limitation—not considered in these designs.

Preliminary design nozzle loads have been computed for a chamber pressure of 800 psi. These are presented as "axial load in nozzle vs. nozzle radius" in Fig. 1 and "nozzle pressure normal to wall vs. nozzle radius" in Fig. 2. It should be stated that only symmetrical loads have been considered in this paper. Preliminary fluid dynamic and thermodynamic analyses were performed to determine the temperature distribution in uncooled nozzle walls of varied design. The predicted temperature distributions in the nozzle walls of each of the structural concepts considered are shown in Figs. 3 — 6. The temperatures and temperature differences shown

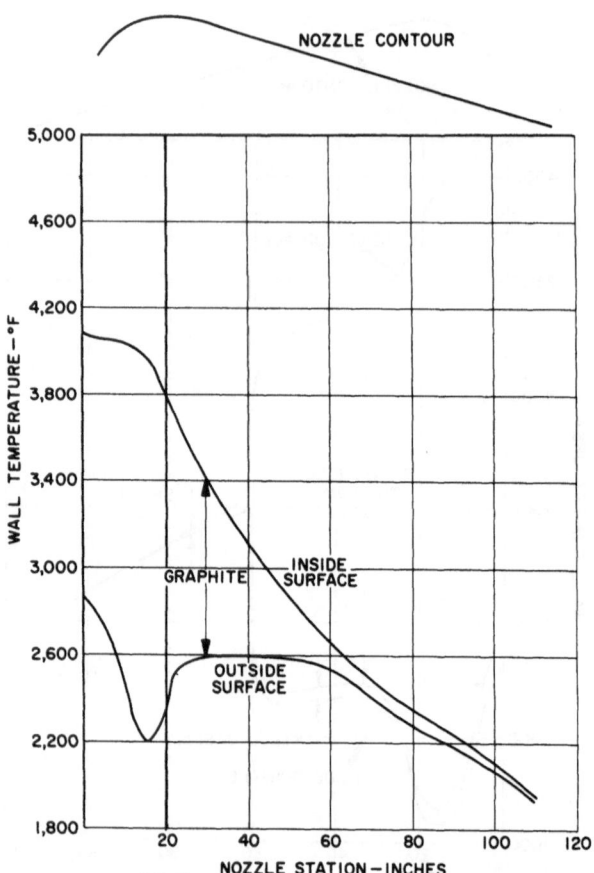

Fig. 5. Predicted temperature distribution in the nozzle walls — concept C.

in these curves illustrate the several problems that must be considered in the material design. Summarizing, these are:

1. High thermal gradients exist which will create high thermal stresses. Mechanisms for relief of excessive thermal stresses must be built into the system.

2. The temperature attained in some materials greatly reduces the creep resistance and could result in significantly large permanent deformations. The extent and effect of these deformations must be evaluated.

3. The temperature attained in the organic–inorganic composites (such as fiberglass), if used, is in excess of that at which physical decomposition occurs.

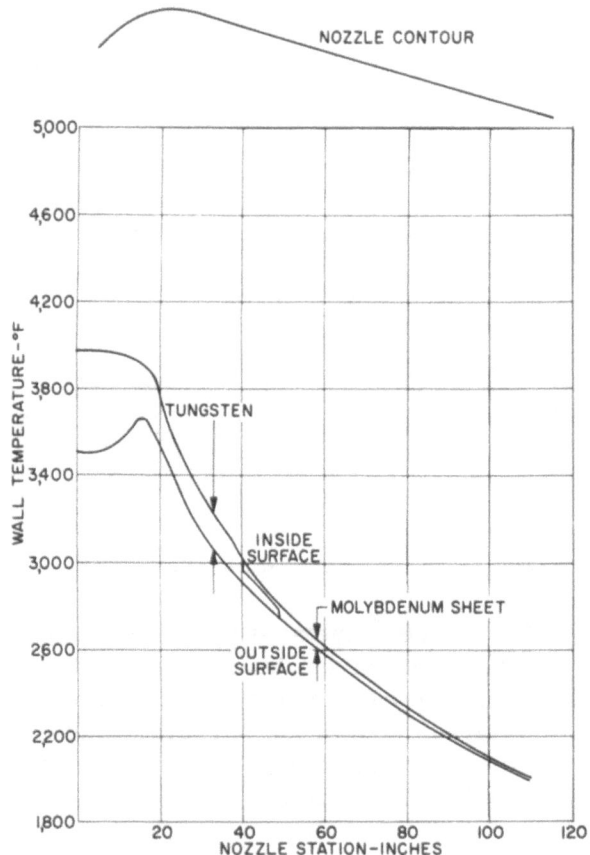

Fig. 6. Predicted temperature distribution in the nozzle walls — concept D.

Based on the above considerations, four material design concepts are presented and grouped according to the type of heat
transfer desired throughout the nozzle material system. A 4000°F
hydrogen gas flow for 30 min and longer is assumed for all design
concepts. The systems are illustrated in Figs. 7—10.

System A—Insulated Structure—Graphite Support

In this system, graphite provides the principal nozzle shape,
and supports the tungsten hot hydrogen gas erosion barrier. In
order to prevent failure of the graphite due to tension cracking,
this component is insulated and wrapped with a refractory metal
strip. By wrapping the outer periphery of the insulated graphite
in tension, one can produce a constant compression stress in the

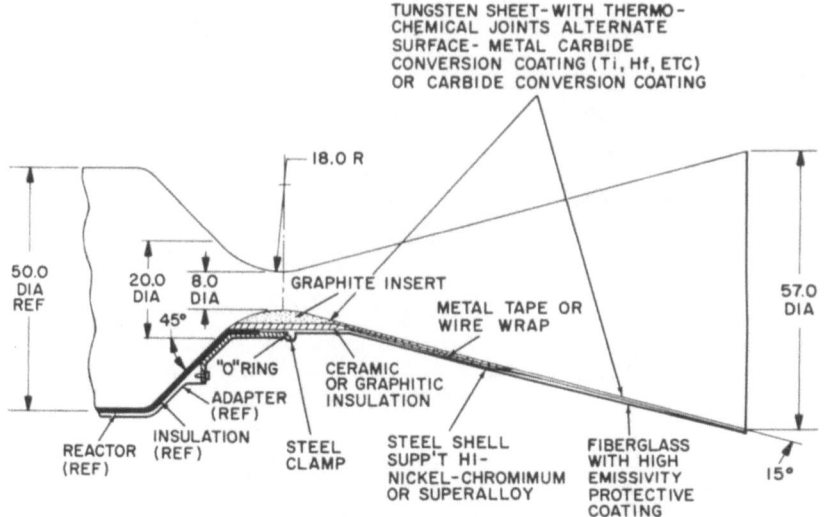

Fig. 7. Nozzle design A.

graphite. The choice of metal strip, of course, is dependent upon the type and thickness of insulation between the graphite and the strip. In order to reduce the temperature of the nozzle assembly outer shell to a sufficiently low temperature to ensure structural integrity, insulation (ceramic or graphic) can be placed around the outer surface of the graphite. At the lower-temperature, exit-end of the nozzle, fiber-reinforced phenolic insulation may be used

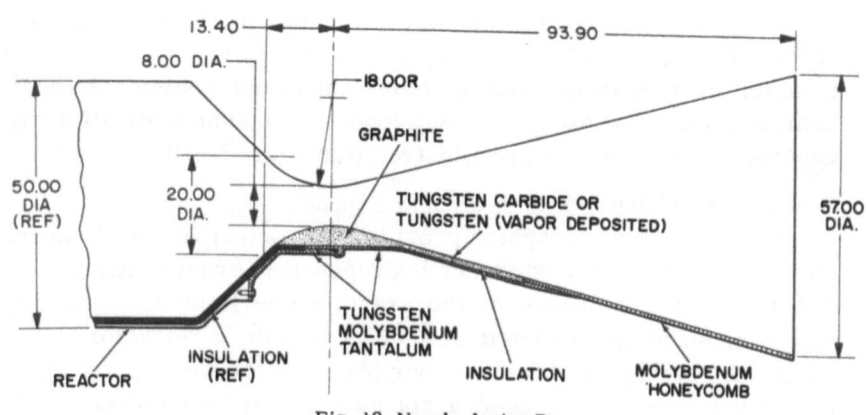

Fig. 10. Nozzle design D.

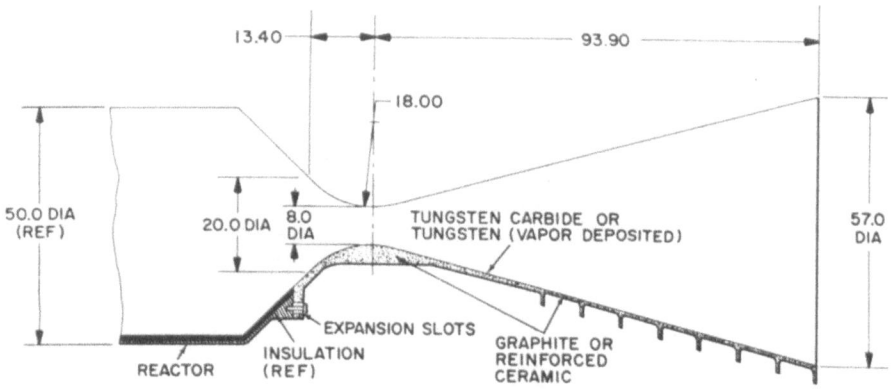

Fig. 9. Nozzle design C.

with a high-emissivity protective coating.   A steel or superalloy outer shell supports the nozzle system.

Estimated temperatures at various locations within this system are indicated in Fig. 3.

Erosion Barrier—Joined Tungsten Sheet. A promising throat insert material for long-time resistance to hot hydrogen gas at 4000°F would be thin tungsten sheet. Tungsten's objectionable factors are its thermal shock behavior, poor fabricability, and high density. In order to minimize these factors, thin sheets of the metal could be hot-formed or roll-formed into the desired nozzle contour shape. The use of thin sheets of 0.020—0.040-in. thickness has shown considerable promise in initial tests [2]. The sections of sheet then are joined by vapor deposition or electron beam welding.

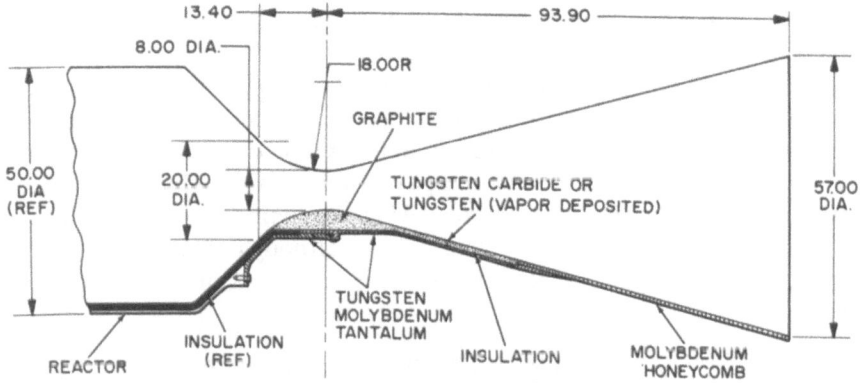

Fig. 10. Nozzle design D.

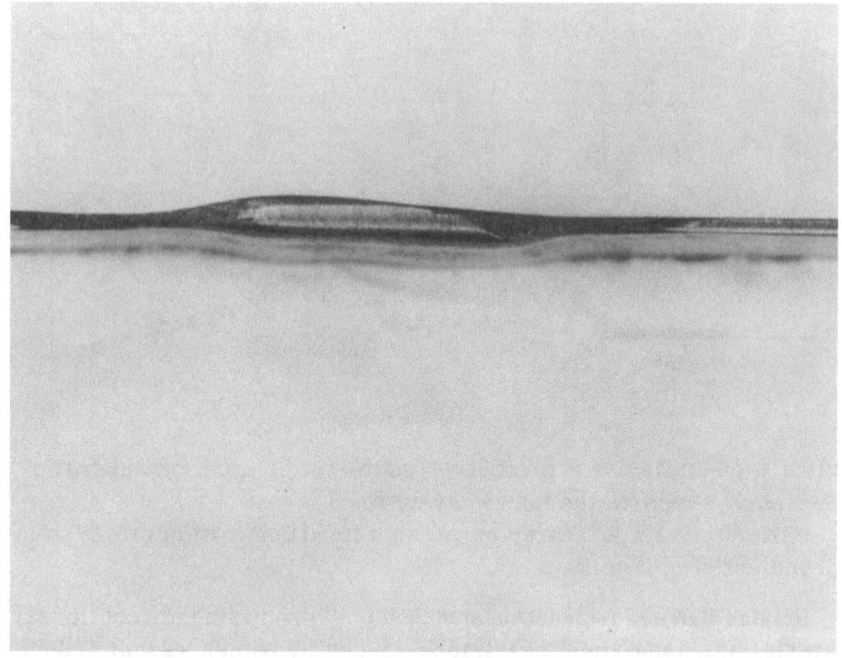

Fig. 11. Edge view of 0.020-in.-thick butt joint of tungsten sheet of thermochemical reduction.

Joining refractory metals by thermochemical vapor deposition has been successfully demonstrated in a recently completed Martin preliminary study [3]. Joints produced by this process are similar to welded joints, but are not susceptible to the deleterious effects of recrystallization caused by current welding techniques. Figure 11 illustrates an edge view of typical butt joint in 0.020-in.-thick tungsten sheet without any significant change to the base metal. The term "joining" describes the new method, because bonds are formed at temperatures far below the level required for fusion and electron-beam welding (where fusion is achieved at the melting point of the base metal). Figure 12 compares an electron-beam-welded 0.020-in.-thick sheet of tungsten with the same sheet joined by thermochemical deposition. The joints, although similar to those obtained by welding, do not affect the grain structure of the base metal.

Support Structure—Graphite.   The use of graphite as a high-temperature material for rocket nozzles has been under investigation

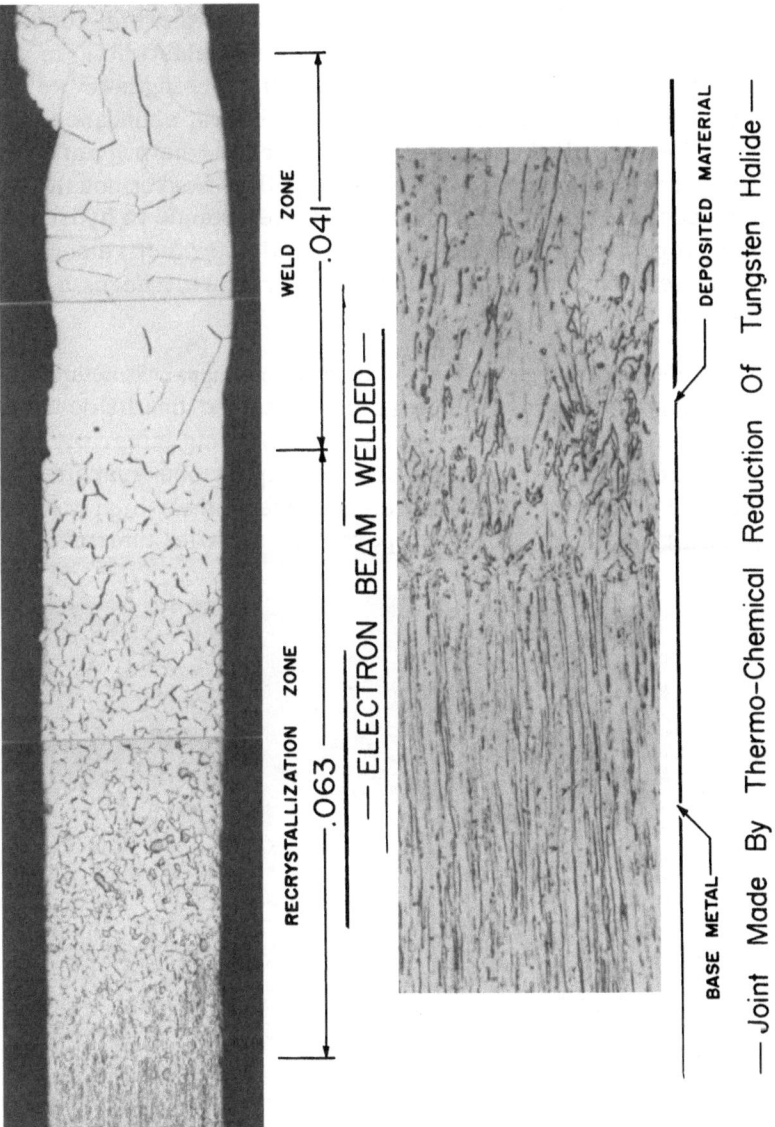

Fig. 12. Microstructural comparison of two methods of joining 0.020-in.-tungsten sheet.

throughout the industry since the advent of rocketry. The prime advantage of graphite lies in its excellent resistance to thermal shock, which is attributed to the material's high thermal conductivity, low thermal expansion, and low modulus of elasticity. In addition, graphite is desirable because of the high strength-to-weight ratio, ease of machineability into complex shapes, abundance, and low cost. Despite all these excellent properties, the utilization of graphite is presently limited to solid-propellant rocket nozzle applications because of the materials' poor resistance to hydrogen. Until this property of graphite can be improved, a protective coating, such as the refractory tungsten liner clad to graphite, affords a possible solution.

Interface Between Tungsten Erosion Barrier and Support Structure. Initial research efforts in 1958 were concentrated on the development of suitable brazing filler metals for producing integrally clad refractory metals to impervious and nonimpervious graphite [4]. Titanium and zirconium were chosen because of their high melting points, ability to form high-temperature, ductile intermetallic

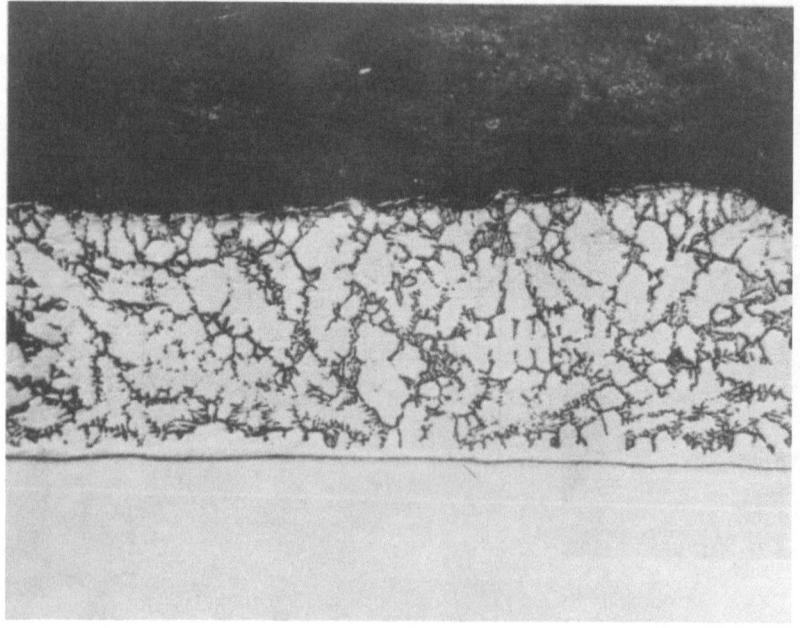

Fig. 13. Tungsten—titanium—TiC (interface) graphite - 50×, unetched.

compounds with the refractory metals and interfacial carbides with graphite, excellent wettability for both the metals and graphite, availability, and cost.

Metallographic examination of the titanium-brazed specimens revealed excellent wetting of the tungsten and graphite, and good bonding except for varying degrees of porosity in the bond layer, satisfactory edge-filleting, and the formation and penetration of carbides at the graphite–titanium interface. Figure 13 shows a typical microstructure of a titanium-brazed specimen. Specimens brazed with zirconium revealed that the carbide formation and penetration into the graphite layer occurred in a manner very similar to that of specimens brazed with titanium.

To further evaluate the bond strength of these brazing metals, shear specimens were fabricated using tungsten–graphite brazed with titanium and zirconium. Testing was conducted in accordance with ASTM specifications A-263-44T. This test was modified, however, as the specimens were reduced approximately 50% in size. Shear values obtained were approximately equal for the two brazing metals with results in the 1800–2000 psi range [5]. All specimens failed at the titanium carbide–graphite and zirconium carbide–graphite interfaces, which indicated that the strengths of the brazing bonds equal or excel the strength of the graphite in shear at ambient temperatures.

A cursory design examination of the structural graphite nozzle coated internally with tungsten sheet was made. If we assume an operating life of 100 hr, the graphite nozzle would have to be only 0.57 in. thick at the throat and taper to only 0.08 in. thick at the station corresponding to a nozzle radius of 8 in. Tungsten gauges woud be 0.72 and 0.074 in., respectively.

There are other design conditions that may prove more critical. Shock loads either on the ground or in flight will surely dictate a greater thickness of structural material at the throat section. Asymmetrical loads from guidance devices must also be considered. Vibratory loads from acoustic excitation or turbulent flow in the exhaust must be considered in establishing minimum gauges near the end of the nozzle.

A plain tungsten nozzle may be too heavy and costly for serious consideration. Tungsten would require the same thickness as graphite to carry the exhaust pressure; however, it would weigh 11 times as much.

System B—Insulated Structure-Reinforced Ceramic Support

This system considers the same type of wrought refractory metal erosion barrier material as described in System A; however, an insulating, refractory metal wire reinforced ceramic replaces graphite. Reinforced ceramic is used because it is a material with thermal expansion more compatible with the support structure than is graphite, and because it reduces the heat transfer to the outer shells. The relative thermal expansion of selected refractory materials is shown in Table I. The ceramic affords an appreciable thermal drop between the inner nozzle and the metallic substrate, so that the substrate can perform its structural support function without a weakening thermal influence. By selecting the proper reinforced ceramic, a thinner material system can be used and the backup insulation reduced, possibly leading to an overall lighter weight system. The thickness ratio of ceramic-to-metal can be predetermined by thermodynamic and thermochemical analysis.

In order to develop the reinforced ceramic structure as an effective thermal barrier to the temperatures developed in an uncooled nuclear rocket nozzle, an optimum combination of refractory metal reinforcement and ceramic matrix is required that effectively insulates and yet provides sufficient metal to overcome the inherently poor mechanical properties common to solid ceramic bodies. Past experience, both in the laboratory and on full-scale solid rocket motors, has reliably demonstrated that refractory ceramic bodies can be made to withstand severe mechanical and thermal stresses by the introduction of a metal reinforcement into the ceramic matrix [6].

Reinforced ceramic structures, with metallic reinforcement means imbedded therein, have been extensively tested by the Martin Company. The reinforcement serves to strengthen the ceramic against vibration and impact shock. It also distributes thermal impact laterally and in depth, making the ceramic much more resistant to thermal shock as well. An ability to resist fracture under stress to a remarkable degree is thus imparted to the ceramic. Since the nuclear motor nozzle will not be exposed to as severe a thermal shock condition during ignition as occurs with solid propellant motors, there is a good chance that this type of structure will survive.

The reinforcement, all-important to the success of this design concept, will require some further investigation. Refractory tungsten wire or filaments, in addition to presently-used molybdenum wire mesh and expanded metal, and corrugated tantalum strips are being considered. In addition, such variables as configuration, spacing, width, thickness, span, and joining techniques should be investigated to assure that optimum conditions are obtained. The reinforcement of the ceramic with randomly oriented tungsten fibers should also be investigated.

For a surface temperature of 4000°F a ZrO_2 composition has given excellent performance. However, for a very long time at 4000°F a ThO_2 composition may be more compatible with a tungsten facing, as indicated in Table II.

The ceramic matrix could be constructed of carefully sized particles, thoroughly blended with a bonding agent to produce a ramming mix or slip consistency, depending on the system being created. The forming of the reinforced ceramic structure will consist of pressing the ramming mix into the reinforcement or slip, coating it under a slight vacuum and finishing to a uniform, predetermined thickness. The ceramic oxides that are considered for the primary constituent of the ceramic matrix should be those materials that can be used in the indicated temperature range and time based on existing state-of-the-art in the use of refractory ceramics.

Estimated temperatures at various locations within this system are indicated in Fig. 4.

System C–Reinforced Ceramic–Oxide or Graphite Structure

This design concept utilizes the reinforced support structure and flame erosion barrier described in Systems A and B. However, the outside shell support structure is removed and the ceramic or graphite is allowed to carry the full internal and external loads.

The unsupported homogeneous nozzle would be made thick enough to resist all loads. The monocoque design has the advantage of minimum differential thermal expansion problems, although the thicker structure could produce excessive internal thermal stresses.

A special form of carbon, produced by vapor deposition at high temperature and known generally as pyrolytic graphite, has certain inherent properties which immediately suggest its probable merit in an uncooled nuclear rocket nozzle. The high sublimation temperature of carbon is well known. The strength of commercial graphite increases with temperature, and at 4500°F is the highest of any known material. Graphite is extremely stable at high temperatures, being moderately sensitive only to oxidation (which is not a problem in space). Pyrolytic graphite shares all of these desirable properties with other forms of graphite. In addition, this vapor-deposited form of graphite offers very high purity and extremely low porosity (high density).

At a high stress level, appreciable creep will occur in graphite above 4000°F. At the same stress level, and at 4500°F, graphite will elongate 2% in 30 min, 4.5% in 90 min, and will fracture in 160 min, even though its strength at this temperature is 75% greater than at room temperature.

Because of the above considerations, it is obvious that the selection of a materials system depends on the predicted total life requirement and reliability capabilities of the nozzle system.

System D–Alternate Flame Erosion Barrier–Carbide Surface on Graphite

For operation at 4000°F flame temperature and for times of up to 30 min for each motor operation, graphite must be augmented with erosion protection to overcome its principal shortcoming, i.e., high erosion or attack with hydrogen gas at high temperature and at long time operation. Therefore, this system is based on the

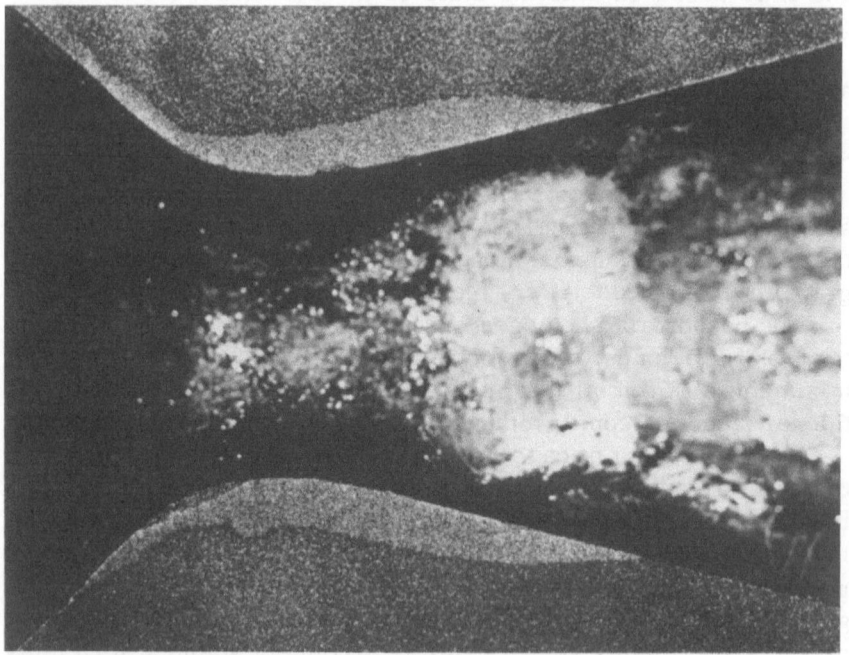

Fig. 14. Cross section of fired rocket (note the diffused SiC coating).

creation of a graphite nozzle that utilizes formed in place mono and binary carbides on the flame surface to improve erosion resistance, and tungsten strip in the outer surface to prevent the occurrence of tensile stress. The formation of stable carbides in the surface of graphite rocket nozzles was developed by Marsh and Locke [7], for the specific purpose of providing graphite with erosion protection. This is illustrated in Fig. 14. Unlike other protective systems for graphite, where the protection compound is deposited on the surface of graphite, the technique used in this concept creates the protective compound in the surface, and then by direct chemical reaction at high temperatures forms the carbide of the diffusing element.

It is important to remember that a detailed analysis is necessary to determine the thermochemistry of the stable carbides and their stability with the reactions of the working gas.

The following discussion only summarizes one of the possible methods of preparing a carbide surface. The methods of preparation presented are the result of Marsh and Locke [8] in their development of carbide-coated solid rocket nozzles. This method considers the direct union of the metal with graphite and the dissociation of a weak gaseous compound of the metal to give finely-divided or nascent metal which can form carbides in the graphite surface. The metal iodides and metal hydrides could serve this purpose. The reaction expressed generally is

$$M + C \rightarrow M_x C + M_{1x} (\text{Free Metal})$$

in which M is tantalum, hafnium, columbium, etc.

Since the free energy change is a measure of the probability of a reaction, this quantity is calculated as a function of temperature. The more negative the free energy, the larger the equilibrium constant according to the reaction

$$\Delta F^\circ = -RT \ln K$$

in which ΔF° is the standard free energy change; R, the gas constant; T, the absolute temperature; and K, the equilibrium constant. In addition, the free energy change is calculated by using the following expression:

$$\Delta F^\circ = \Delta H^\circ - T \Delta S^\circ$$

The heats and entropies of transitions and phase changes of the carbides under consideration are available, so far, for only a few cases. Also heat capacity data appear scarce. Therefore, the results would be only first approximations, since H and S are functions of temperature and heat capacity. This could be illustrated by the following expressions:

$$\Delta H_T = \Delta H_{298} + \int_{298}^{T} \Delta C_p \, dT + L_T + \int_{298}^{T} \Delta C_p \, dT$$

and

$$\Delta S_T = \Delta S_{298} + \int_{298}^{T} \frac{\Delta C_p}{T} \, dT \pm \frac{L_T}{T} + \int_{298}^{T} \frac{\Delta C_p}{T} \, dT$$

It is expected that, in many cases, the changes in ΔH and ΔS would offset each other, and the error would not be large, that is, if a constant ΔH and ΔS are used.

IV. CONCLUSIONS

It is not possible at this time to determine which of the designs is the most promising. However, designs warranting study are the composites where the materials are chosen individually for erosion, insulation, and strength. In addition, such conclusions must await the further evaluation of the results presented herein in terms of the stresses, permanent deformations, and material allowables that occur at the various temperature levels. Also, a great deal of material property data must be obtained at high temperatures. Creep data on some materials in question are virtually nonexistent and must be determined before the necessary thermal analysis can be made.

REFERENCES

1. S. R. Locke, Tungsten Clad Graphite Nozzles, NASA-WADD, High-Temperature Material Conference, San Diego, California, January, 1960.
2. Vector Control Report on Static Firing of Engine, Aerojet-General Corp. ICM, from L. I. Barsky to R. S. Newman, dated September 12, 1960.
3. S. R. Locke and J. Macedo, Thermochemical joining, patent applied for, Martin Company, 1962.
4. S. R. Locke, The Brazing of Refractory Metals to Graphite for Solid Rocket Nozzles, Special Report, Aerojet-General Corp., May, 1959.
5. S. R. Locke, ibid.
6. A. H. Levy, S. R. Locke, and H. Legett, Composite Ceramic—Metal Systems for 3000°F to 6000°F Service, Astronautics, April, 1961.
7. L. Marsh and S. R. Locke, Carbide Coated Graphite Nozzle for Solid-Propellant Rockets, JANAF Proceedings, Vol. III, June, 1959.
8. L. Marsh and S. R. Locke, ibid.

AUTHOR INDEX

SUBJECT INDEX